USS SALEM CA-139

WARSHIP PICTORIAL #8

CLASSIC WARSHIPS PUBLISHING

Editor/Research - Steve Wiper
Layout/Illustrations - T.A.Flowers

Current titles in print from
CLASSIC WARSHIPS PUBLISHING

USS INDIANAPOLIS CA-35
USS MINNEAPOLIS CA-36
USS LOUISVILLE CA-28
USS TEXAS BB-35
USS SAN FRANCISCO CA-38
OMAHA CLASS CRUISERS
NEW ORLEANS CLASS CRUISERS

CLASSIC WARSHIPS PUBLISHING
P.O.Box 57591 • Tucson AZ 85732
ISBN #0-9654829-7-9

Look for these exciting titles in future books

Yorktown Class Aircraft Carriers
Sims/Benson/Gleaves Class Destroyers
USS Nevada BB-36
Japanese Battleships of WWII

USS Enterprise CV-6

USS Salem CA-139, *sometime in the early 1950s. US "All Gun" cruiser design had reached its climax in the* Des Moines *class. They were the accumulation of hull design, weapons technology, and featured the cutting edge in radar and electronic countermeasures. They were capable of the same main battery fire power of three of their predecessors, and could out perform any of the older battleships, prior to the* South Dakota *class.*

USS Salem CA-139

Des Moines Class Cruiser

Operational history

1943

June 14 - *CA-139* was ordered by the US Navy. Many design changes occurred thru this year and 1944, before a final configuration was developed.

1945

July 4 - The keel was laid down at the Bethlehem Steel Company's Quincy Yard in Quincy, Massachusetts.

1947

March 25 - *CA-139* was launched, sponsored by Ms. M. G. Coffey.

1949

May 14 - She was commissioned as *USS Salem*, at the Boston Navy Yard, under the command of Captain J. C. Daniel.

June - Visited namesake port, Salem, Massachusetts.

July 4 - Departed Salem, MA. for a four month shakedown cruise, back and forth, between Boston and Guantanamo Bay, Cuba, earning the nickname "Gitmo Express".

October - Received post shakedown cruise repairs at the Boston Navy Yard.

November/December - Sortied two more cruises in the Caribbean to Guantanamo Bay, Cuba.

1950

January/March - *Salem* participated in maneuvers with the US Atlantic Fleet in both the Caribbean and Atlantic Oceans. Command was transferred to Capt. E. B. Taylor on February 2nd.

May 3 - Departed Boston for the Mediterranean Ocean, to relieve *USS Newport News CA-148*, as flagship of the US 6th Fleet, on the 17th.

June/August - Visited ports in Malta, Italy, Greece, Turkey, Lebanon, and Algeria, and participated in maneuvers and training exercises.

September 22 - Relieved by *Newport News* and returned to the United States for resupply and repairs.

October/December - After three weeks of repairs at Boston, *Salem* joined the US Atlantic Fleet for maneuvers.

1951

January/February - She spent six weeks of intense gunnery training off Guantanamo, afterwards, further training off of Bermuda. Command changed during this period to Captain D. C. Varian.

March 20 - Steamed for the Mediterranean to relieve *Newport News* as flagship of the US 6th Fleet. It was during this tour that she was christened "Pride of the Fleet", receiving the Fleet Battle Readiness Pennant for performance in Battle Readiness Competition, within the US Atlantic Fleet.

September 19 - Relieved by her sister ship, *USS Des Moines CA-134*, and returned to the US for four months of refit and overhaul at Boston.

November 9 - Captain W. K. Romoser took over command.

1952

February/March - After repairs, a refresher training cruise to Guantanamo and back to Boston for minor repairs.

April 19 - *Salem* steamed for the Mediterranean for her third deployment, relieving *Newport News* as flagship, at Algiers on April 28.

May/September - Aside from the normal port of calls and exercises during this time, she participated with units of other Allied Navies in "Exercise Beehive II". The navies included were British, Italian, French, and Greek.

September 29 - Relieved by *Des Moines*, returned to the US, arriving at the Boston Navy Yard on October 9.

October 9 - Captain B. Schumm, assumed command, operated *Salem* off the US eastern seaboard for four months.

1953

January 24 - Departed for Guantanamo Bay for training exercises. On the 28th, Captain L. W. Creighton assumed command. Returns to Boston on February 27.

April 17 - Departed for the Mediterranean, again to relieve the *Newport News* as flagship of US 6th Fleet. On this, her fourth deployment to the Mediterranean, *Salem* participated in "Exercise Weldfest". She was also the first ship on scene in the earthquake ravaged Greek Ionian Islands, where she would assist relief and give humanitarian aid until her supplies ran low, four days later. The *Salem* and her crew earned the praise of the King and Queen of Greece for their efforts.

October 9 - Relieved by *Des Moines*, she returned to Boston on the 24th, for an extensive refit and overhaul.

1954

February 6 - Operated off Cuba for refresher training until April 7 and returned to the Boston Navy Yard, to prepare for her next deployment.

April 30 - Steamed for the Mediterranean Ocean to relieve *Newport News* as the US 6th Fleet Flagship, on May 12.

September 22 - Relieved by *Des Moines* and returned to Boston on the 29th.

October/December - Participated in US Atlantic Fleet exercises off the US

Eastern Seaboard, and the Caribbean Ocean. The *Salem* received a small refit and overhaul at the Boston Navy Yard at the end of the year.

1955

January 14 - Captain J. Maginnis took command of *CA-139*.
January 19/February 22 - Undertook annual journey to Guantanamo Bay for gunnery drills, refresher training, and a naval reserve training cruise.
May 2 - Steamed for the Mediterranean on her sixth deployment with the US 6th Fleet, relieving the *Newport News,* as flagship, on May 19. Over the next five months *Salem* participated in a NATO (North Atlantic Treaty Organization) exercise, and a Franco-American naval exercise. During this last operation, she embarked the US Under Secretary of the Navy, T. S. Gates, as an observer.
September 23 - Departed Barcelona, Spain, for Boston Navy Yard, arriving October 2, for a four month refit and overhaul.

1956

January 14 - Captain A. Roby assumed command.
January 16/April 5 - Steamed on refresher training and gunnery practice cruise to Guantanamo Bay and returned to Boston Navy Yard to resupply.
May 1 - Steamed for the Mediterranean for her seventh deployment as the US 6th Fleet Flagship. While en route, the "Suez Canal Crisis" broke out.
May 14 - Diverted to the eastern Mediterranean Greek island of Rhodes, she joined the fleet as flagship.
June - *Salem* remained in the eastern Mediterranean Ocean until the tensions subsided, returning to her normal operations out of Villafrance, her homeport for a 20 month deployment as flagship.
October 30 - Returned to the Suez Canal area when fighting broke out to assure the safe passage of American and other friendly ships thru the canal zone. Her presence needed, she remained in these waters for the most part, thru the beginning of the following year.

1957

February 15 - Captain F. T. Williamson became the *Salem's* next commanding officer.
March - The crew of turret #3 aboard the *USS Salem* won the coveted "Gold E" for excellence in gunnery and outstanding gun crews.
April - Hostilities threatened peace in the canal zone again, forcing a presence by the cruiser and other US 6th Fleet vessels.
May/July - Normal operations resumed, she visited ports throughout the Mediterranean Ocean.
August - Hostilities arose once again, and again *Salem* and other NATO vessels make their presence known in the tumultuous eastern Mediterranean.
October/December - The US 6th Fleet made a show of force to support the government of Jordan, threatened by subversion, in the eastern Mediterranean.

1958

January/June - Remained flagship of the US 6th Fleet, the *Salem* made her and the fleet's presence known to all the far reaches of the Mediterranean, still based at Villafrance, on the French Riviera. Her crew enjoyed a good tour of duty in the sunny Mediterranean.
June 26 - The *USS Salem* departed for her return voyage home to the United States, arriving at the Norfolk Navy Yard on July 4. The ship was scheduled for inactivation into the "Mothball Fleet".
July 23 - Captain J. D. Ferguson assumed command. The cruiser received a short reprieve when the government of Lebanon requested assistance against an anticipated coup on this date.
August 11 - *Salem* relieved *USS Northampton* CLC-1 as Commander, US Second Fleet.
September 2 - Departed Norfolk for the Mediterranean, arriving at Augusta Bay, and Barcelona, Spain, for a ten day emergency cruise.
September 30 - Arrived back at Norfolk Navy Yard, disembarking crew and off loaded ammunition and supplies in preparation for inactivation.
October 7 - Reported for inactivation.
October 25 - Commander of the US Second Fleet disembarked.

1959

January 30 - Decommissioned and moved up to the Atlantic Reserve Fleet at the Philadelphia Navy Yard.

1994

October 30 - The *Salem* made her way north to the place of her origin, the Bethlehem Steel Shipyard, in Quincy, Massachusetts. The ship was made the centerpiece for the United States Naval & Shipbuilding Museum.

1995

May 14 - *Salem* was recommissioned, 46 years to the day of her original commissioning, into the Historic Naval Ships Association.

1996

July 2/7 - Participated in the *USS Constitution* "Turn Around" in Boston Harbor with other museum ships.

1999/2000

December 7/February 23 - The *Salem* was towed to the Boston Ship Repair Yard and dry docked for hull scrapping and re-painting.

The *USS Salem CA-139,* at the date of this printing, is the only United States Navy heavy cruiser museum. If your in the Boston area, visit a unique part of American military history.

USS Salem CA-139 *is about to be launched, March 27, 1947. This new class of heavy cruiser was the biggest yet to date.* Salem *and her sisters,* Des Moines CA-134 *and* Newport News CA-148, *were a full 43 feet longer than the previous* Baltimore *class cruisers. The* Des Moines *class beam was over 4 feet wider. Their displacement was almost 4,000 tons greater. The general shape and look of* Salem's *hull was simular to the earlier class, but was larger and had slightly exaggerated features. Looking closely, along the side of the ship, the transition in hull form caused by the prominent knuckle can be seen. The large bulbous bow would help with buoyancy, keeping the bow from plunging into a heavy sea. Note the bow numbers at this date are black on the light, Haze Gray hull. The tall string of numbers running up the forward edge of the bow are draft markings which are also located on the stern. These will give the depth of the hull from the standard load waterline.*

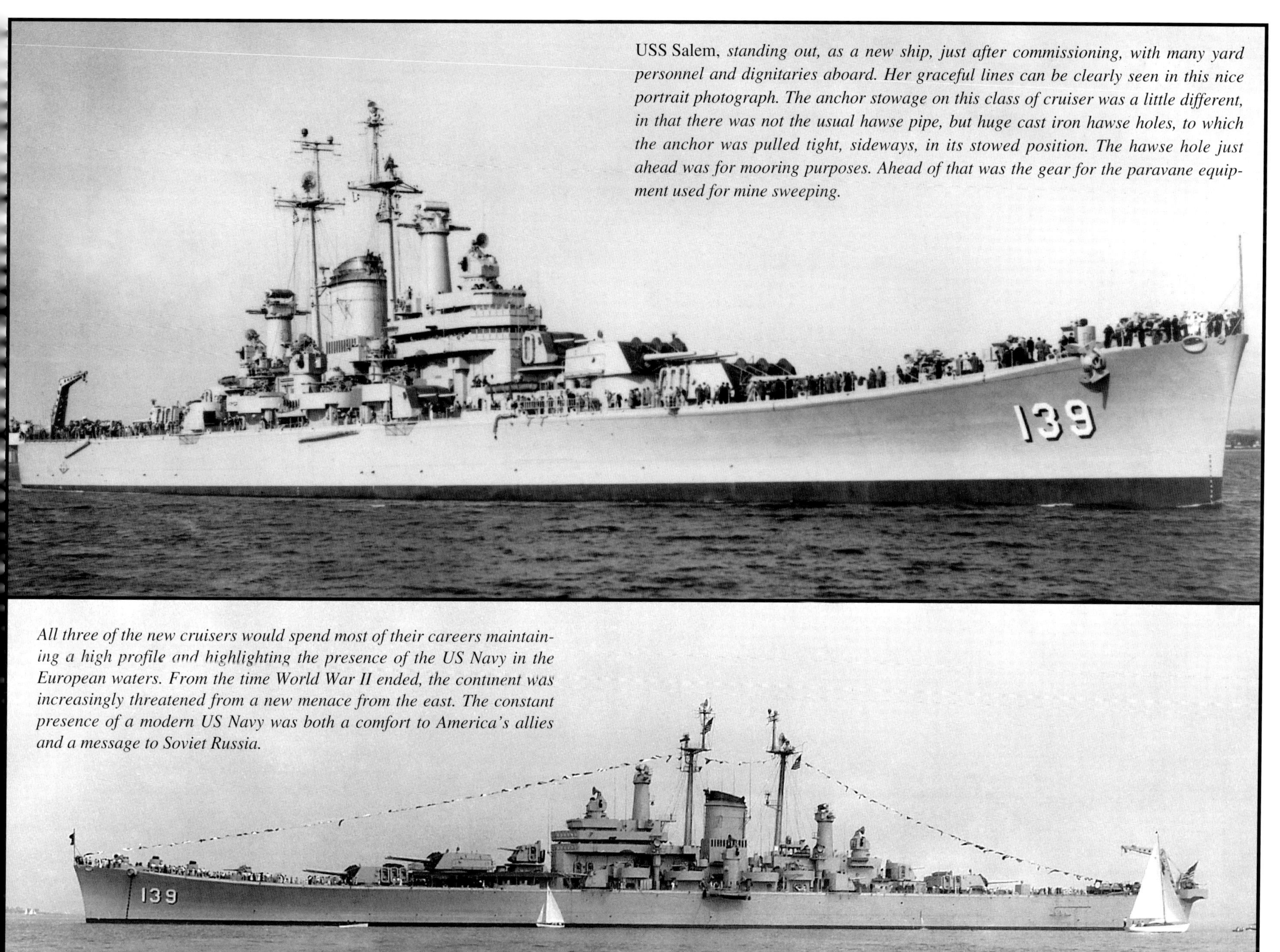

USS Salem, *standing out, as a new ship, just after commissioning, with many yard personnel and dignitaries aboard. Her graceful lines can be clearly seen in this nice portrait photograph. The anchor stowage on this class of cruiser was a little different, in that there was not the usual hawse pipe, but huge cast iron hawse holes, to which the anchor was pulled tight, sideways, in its stowed position. The hawse hole just ahead was for mooring purposes. Ahead of that was the gear for the paravane equipment used for mine sweeping.*

All three of the new cruisers would spend most of their careers maintaining a high profile and highlighting the presence of the US Navy in the European waters. From the time World War II ended, the continent was increasingly threatened from a new menace from the east. The constant presence of a modern US Navy was both a comfort to America's allies and a message to Soviet Russia.

This photograph was taken sometime near her commissioning date, possibly upon her completion, May 9, 1949. It shows an interesting view of the bridge structure, funnel and mastwork. The long rectangular superstructure design enabled the large amounts of anti-aircraft batteries to have a maximum arc of fire. The panels at the lower edge of the windshields on both bridge levels were for splinter protection, as they could be raised up to cover the glass panels. On the uppermost bridge level (Fire Control), the baffle running along the forward edge is a wind deflector, designed to direct airflow up and over the bridge personnel. The rounded main deck edge is clearly seen here, which is a carry over from the previous Baltimore *class cruisers. At this early date in her career,* Salem *still retains the 25 man rectangular life rafts and floater nets, later to be replaced with inflatable boats. The two masts were massive, supporting an abundance of radar and radio antenna, and were a distinctive feature of this class of ships.*

These views allow us to see the general layout of this class of ships. The tubs for the twin 3in./50cal. automatic anti-aircraft guns at this time have a fully rounded and extended shield. These were numbered, starting from the bow, working aft, as 31, 32, etc., with the last being 312. They were labeled as such to signify 3in, mount one, two, etc. On the mounts amidships, the even numbers were to port (left) and the odd to starboard (right). Mounts 37 and 38 had

8 in/55 caliber gun

Mk 16/0

Bore	8 in
Weight	16.68 tons
Muzzle velocity	2500 f/s
Working pressure	18.2 ton/in^2
Approx life	780 rds
Rate of Fire	6.1 spm
Max range	31,350 yd HC
	30,000 yd AP
Weight projectile	260 lb HC
	335 lb AP
Propellant charge	78 lb

triple mount turret

weight		451 tons
gun elevation		-5 to 41°
elevation speed		8.2°ps
training speed		5°ps
crew		45
armor (in)		
	face	8
	side	2-3.75
	rear	1.5
	top	4

The main armament of the Des Moines *class cruisers was comprised of the new Mk 16 gun, mounted in three triple turrets, retaining the same layout as earlier USN cruiser designs. One of the innovations of this mount was the fact that they could reload at any elevation. This would allow for an impressive volume of fire at 90 rounds per minute, or about 15 tons of shells, for the total battery of nine guns. One would not want to be on the receiving end of that kind of firepower. Armor protection on the turrets themselves was slightly better than that of the proceeding* Baltimore *class, with the same 8 in face, but an additional inch on both the roof and the sides. There were two noticeable items about the gun turrets in this class that were different from all previous USN designs. The first was the lack of turret mounted range finders. That equipment was limited to the main battery range finders. The other difference was the downward slope to the roof of the turret. From midway, back, the turret roof had a slight downward slope, the opposite of all earlier designs. The small radar antenna in this photo on the roof of the turret, is for Mk27 ranging radar, a back-up fire control system. These antenna were found in pairs on both the #2 and #3 turrets.*

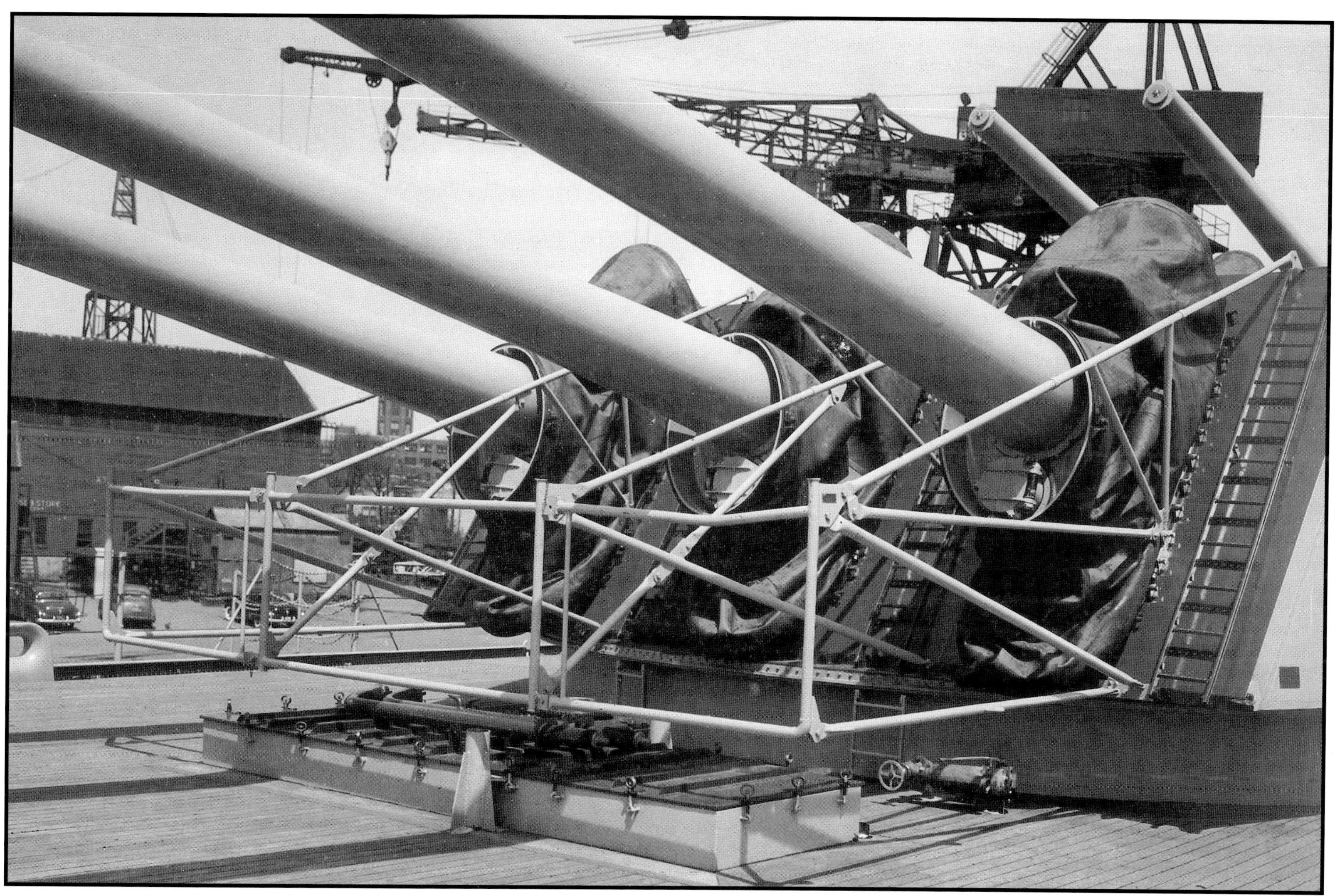

With the netting installed, not pictured here, the large basket mounted on the face of each turret, would catch the forward ejected spent gunpowder casings. The doors to the ejection chutes are visible here on the lower face of the bloomers, just below the gun barrel itself. The spent casings were made of brass and recycled. The bloomers were placed around the gun barrels, and attached to the turret face to weather proof the mount. Notice how all deck equipment in the vicinity of the mounts must be low enough to clear the transverse arc of the turret.

These two views, photo and illustration, give us a better understanding of the main gun turrets. In the photograph, another view of the basket and the overall shape of the turret is visible. The Mk 27 radar antenna is visible here, atop the turret.

The illustration shows the backside of the turret, where the ammunition for the main magazine is resupplied. The small davits mounted on the back are for hoisting the heavy shells, and brass gunpowder casings, up to the turret, and inside. The ammunition is loaded onto the auto feeder, which is run in reverse, carrying the shells down to the magazine. It must have been a real chore to do this, in comparison to the rate at which the ammunition was expended. A "Special Weapons" magazine was located below decks, adjacent to the #2 turret, which was used for storage of chemical, biological, and tactical nuclear 8in shells. This information may never be declassified whether or not any of these weapons were ever aboard. However, t here was an armed Marine guard posted on a 24 hour watch at the magazine entrance.

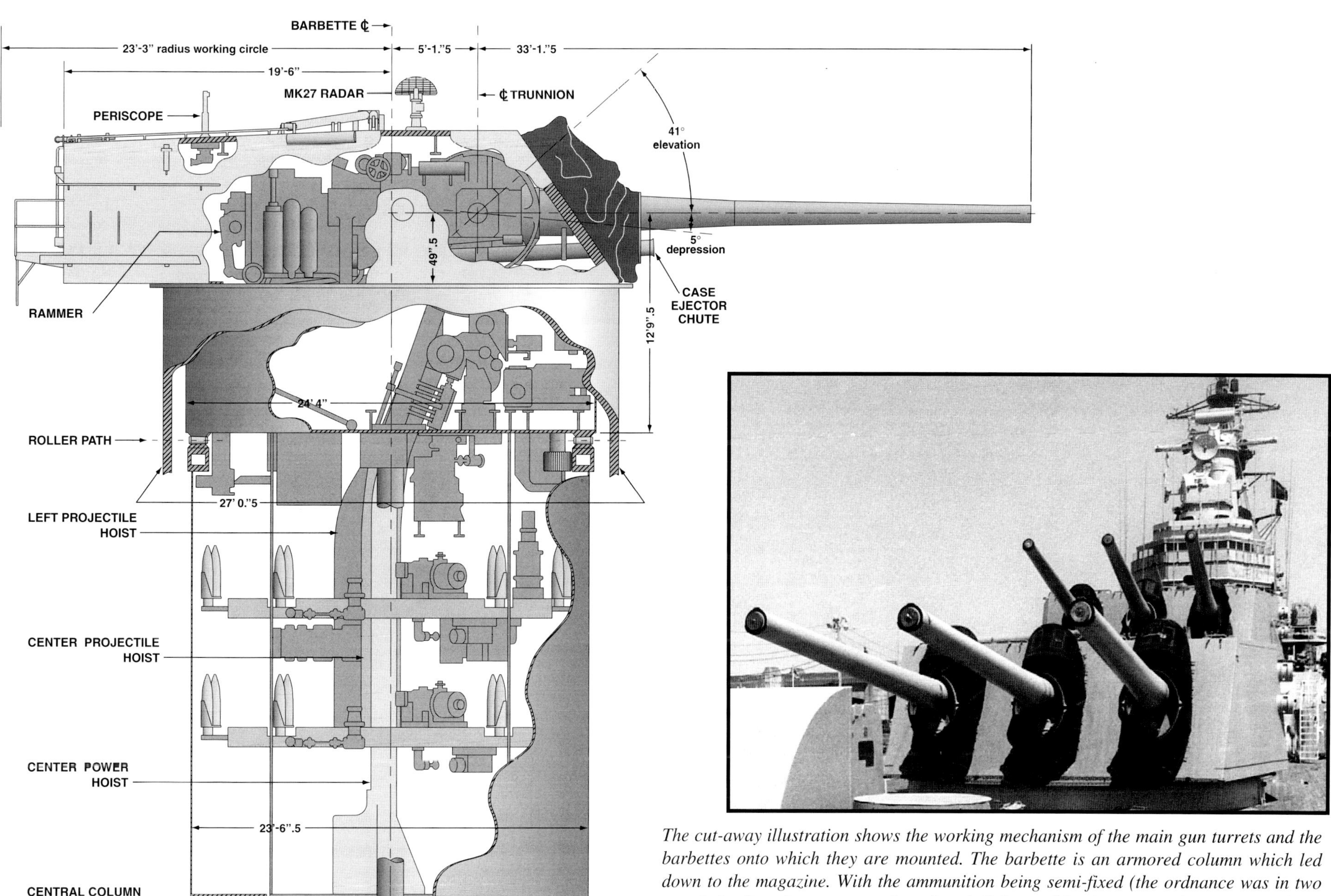

The cut-away illustration shows the working mechanism of the main gun turrets and the barbettes onto which they are mounted. The barbette is an armored column which led down to the magazine. With the ammunition being semi-fixed (the ordnance was in two parts), loading time was greatly reduced. Prior to this, the gunpowder was in multiple silk bags and loaded individually. The procedure took longer and was not nearly as safe. The brass casing was ejected by air pressure, forward thru the ejection chutes described earlier and pictured here.

illustration scale 1/96

Mk54 - 8in director

The Mk54 dual purpose director was the replacement for the Mk34 single purpose director and was designed for the three ships of the *Des Moines* class exclusively. The unit was similar to the Mk38 found on battleships. Pointing and training optics were stabilized by a direct link to the 26.5ft rangefinder. The maximum accurate capability for the director was 45 knots on a surface target and 800 knots on an aircraft. The entire system weighed 20.23 tons and required an operating crew of 30 men.

Mounted atop the director, was the Mk13 radar antenna which first saw use during the later stages of the Second World War. It was an upgrade from the the Mk8 radar used on previous USN vessels. Accurate ranges were 40,000yds on a battleship, 31,000 on a destroyer, 30,000 on a bomber and 10,000 on a submarine (surfaced). The Mk13 was the last main battery director radar to be produced for the US Navy.

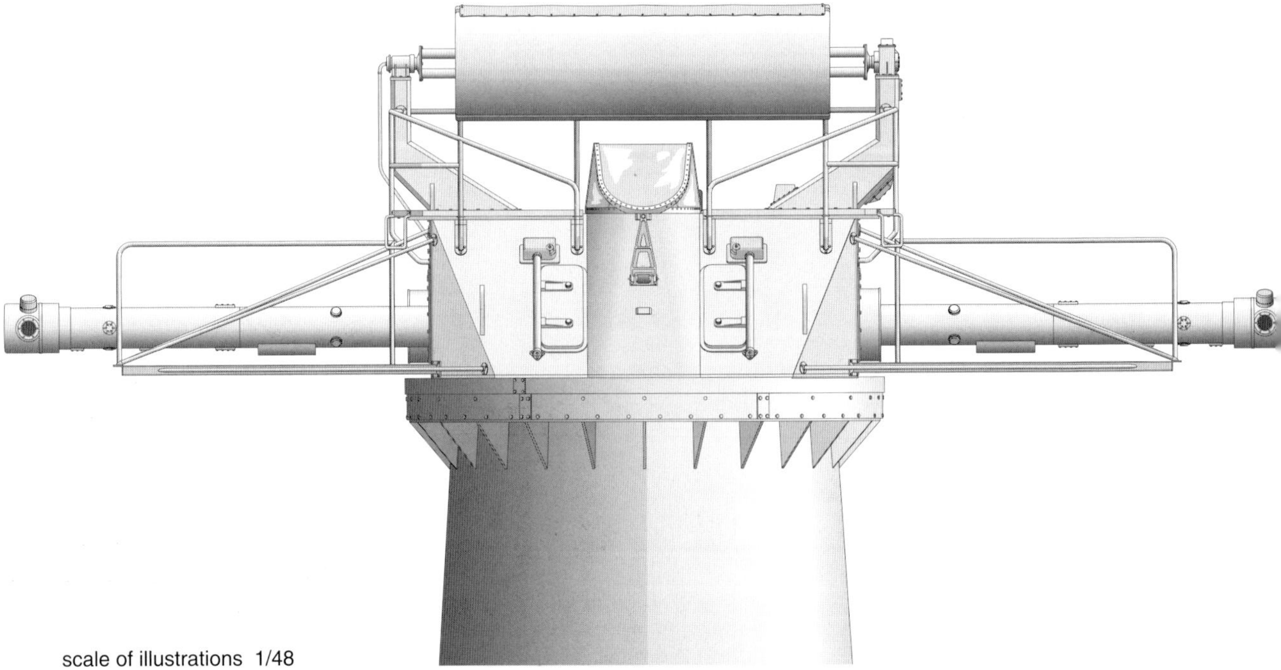

scale of illustrations 1/48

Access to the Mk54 director proper was by means of vertical ladders on the outside of the structure.

USS Salem *and* USS Columbus CA-74 *operating together sometime in the early 1950s, in the Mediterranean. They are probably undergoing a "High Line" transfer, which was where a line was run between the two moving ships, and items and or personnel were transferred across from one ship to another. Needless to say this was a dangerous operation, but it enabled the ships to do so without stoping, saving both time and fuel. This view gives us a nice comparison of the evolution of the* Baltimore *class to the* Des Moines *class cruisers. Notice the differences in the construction and layout of the two superstructures, yet the main, secondary, and anti-aircraft armament arrangement is very simular.*

Date of completion, May 9, 1949.

Operating in the Mediterranean, often as Flagship of the US 6th Fleet, the Salem *earned one of her nicknames, "Pride of the Fleet", which when you look at this photograph, you can see her long graceful lines and bristling guns, made her so. The background setting for both of these photographs really make a great portrait for a classy ship. The harbor is Villafrance, Sur Mer, on the French Riviera, sometime in the early 1950s. She must have at least one of her twin 3in mounts landed for servicing, or repairs, as number 37 appears to be missing. Large canvas tarpaulins have been rigged to keep the hot Mediterranean sun from making the ship to uncomfortable. The darker colored tarpaulins are royal blue in color to signify her as the flagship. The* Salem *did have an air conditioning system of the chilled water type, but it was not as efficient as that aboard her sister,* Newport News, *which had a freon type system. The* Des Moines *did not have one at all. These were the first ships in the USN to be designed with an air conditioning system.*

139

The armored twin 5in. secondary mounts aboard the *Salem* are a dual purpose mounting, meaning they were used against both surface and airborne targets. This was a semi-automatic gun, using fixed ammunition, whose rate of fire was wholly dependent upon the proficiency of the turret crew. These turrets were directed primarily by the four Mk37 directors, situated fore, aft, and amidships, port and starboard. One such director is visible in the photograph below, directly above the 5in turret. The origin of this gun turret starts with the mounts on the *Porter* and *Somers* class destroyers, built in the mid to late 1930s. Progressing thru the secondary turrets aboard the *St. Louis* class cruisers, the first armored units were installed aboard the battleships of the *North Carolina* class. All of the US cruisers and battleships built during World War Two carried this mount as their secondary armament, with the exception of the *Worcester* class cruisers.

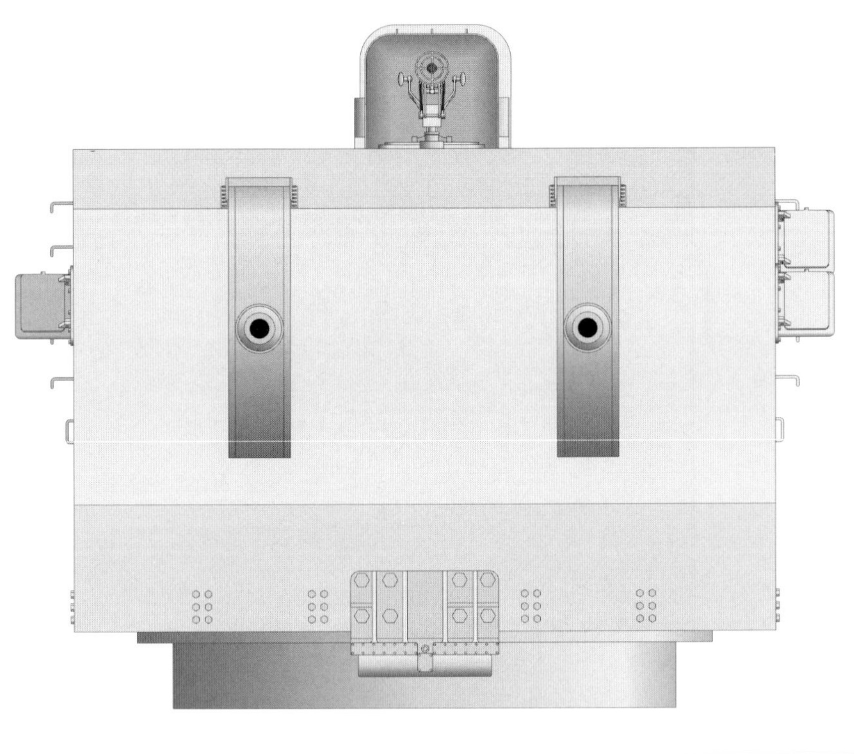

5in SECONDARY BATTERY • DUAL PURPOSE

38cal Mk 12-1

Muzzle velocity	2,600fps
Rate of Fire	20rpm
Maximum range	18,000yd @ 45°
Maximum ceiling	12,400yd @ 80°
Approx life	4,600 shells
Ammunition weight including	15.2lb propellant
anti-aircraft	55lbs
phosphorous	50lb

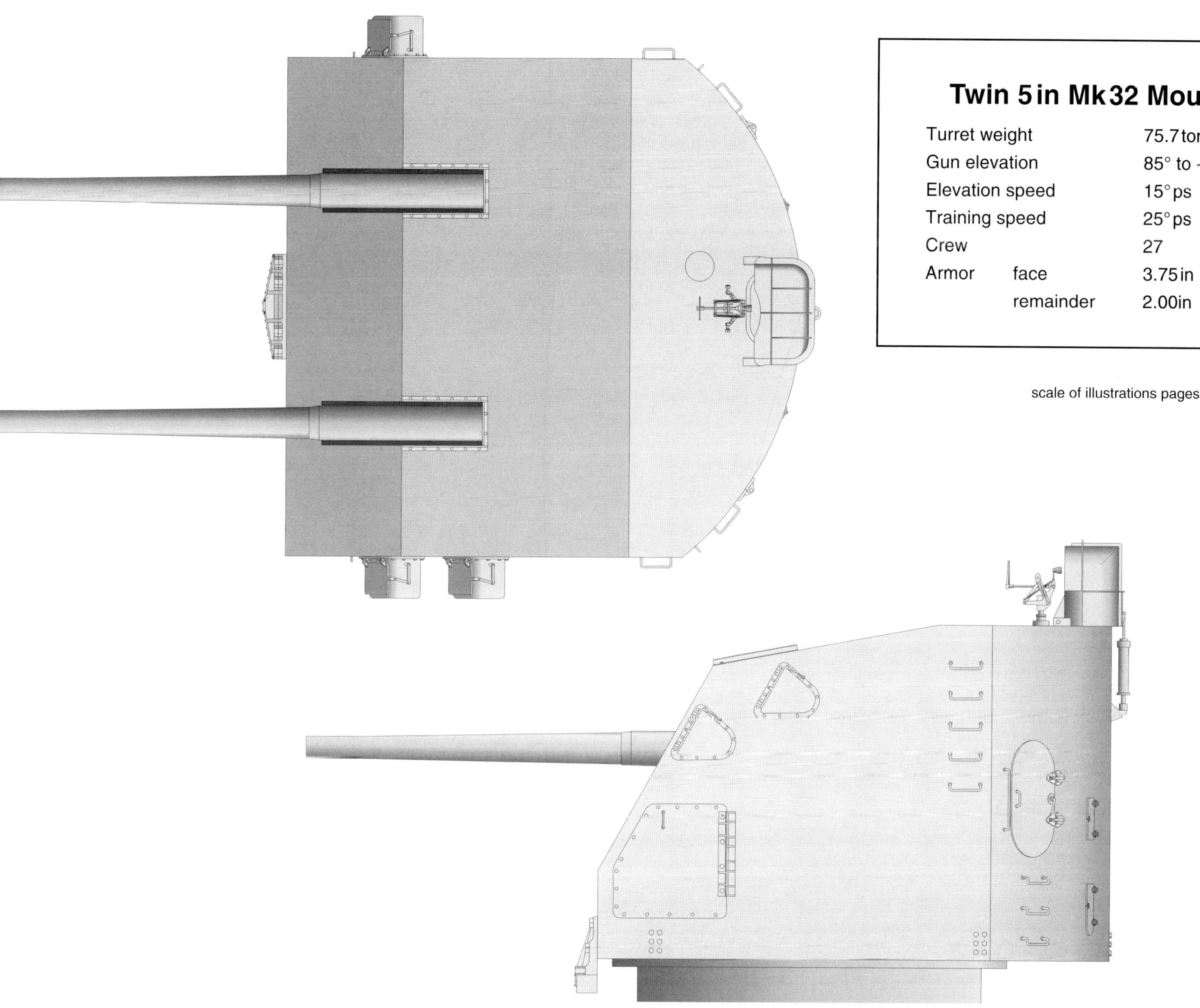

Twin 5 in Mk 32 Mount

Turret weight		75.7 tons
Gun elevation		85° to -15°
Elevation speed		15° ps
Training speed		25° ps
Crew		27
Armor	face	3.75 in
	remainder	2.00in

scale of illustrations pages 20/21 1/48

Mk37 - 5in gun director

The Mk37 was designed in 1936 and originally mounted in the *North Carolina* class battleships and the *Sims* class destroyers. The director was dual purpose and could handle fire control for the main battery as well as the 5in. It carried features like the slew sight and an automatic rate control computer which allowed for fast target solutions. The 15ft rangefinder was stereoscopic and the director had a crew of 6.

Mk25 Fire Control Radar

Mounted above the director was the Mk25 radar antenna. The Mk25 was primarily intended to be used for 5in fire control, but could also be used for fighter control. Fighter control means the control, and or direction, of friendly aircraft, at or near the forward battle area. The final version mounted on the *Salem* was probably Mod3, which was accurate to a distance of 150,000yds. This radar unit was first produced in March, 1945, and as modifications were developed, the final version, as mounted in *Salem*, was even capable of tracking projectiles as they left the gun, to their target.

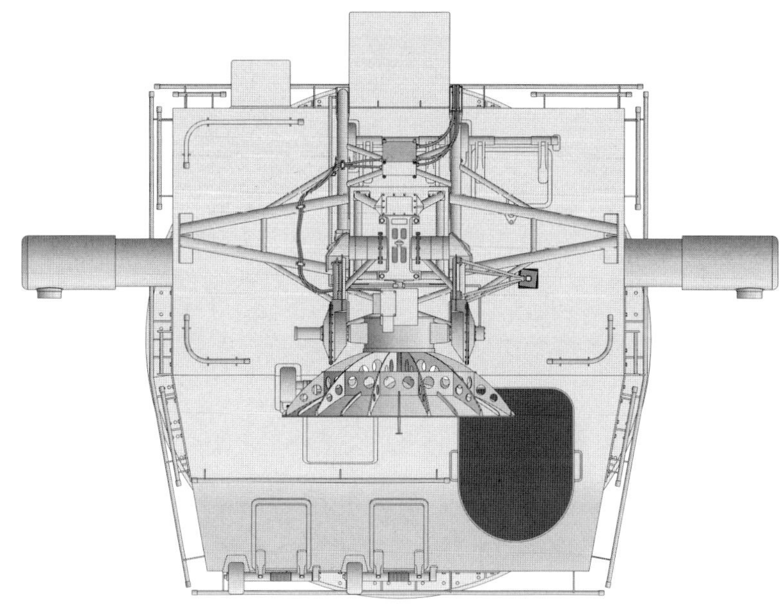

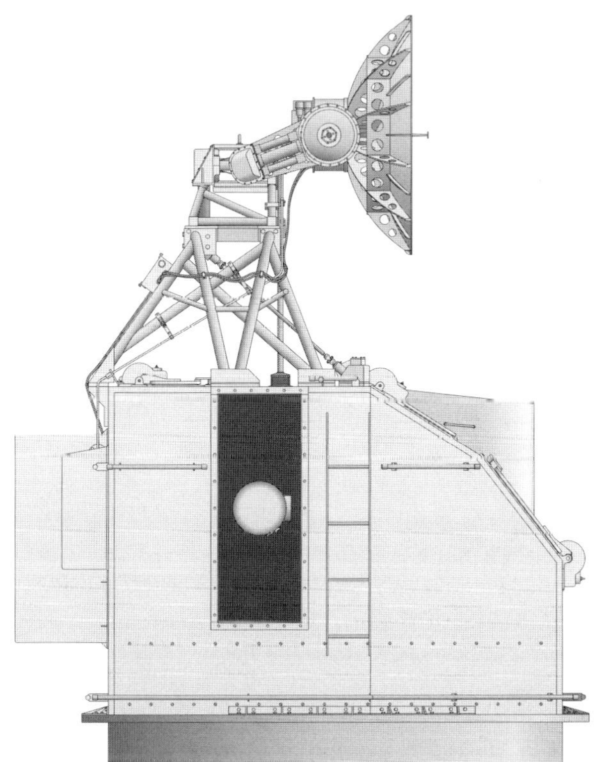

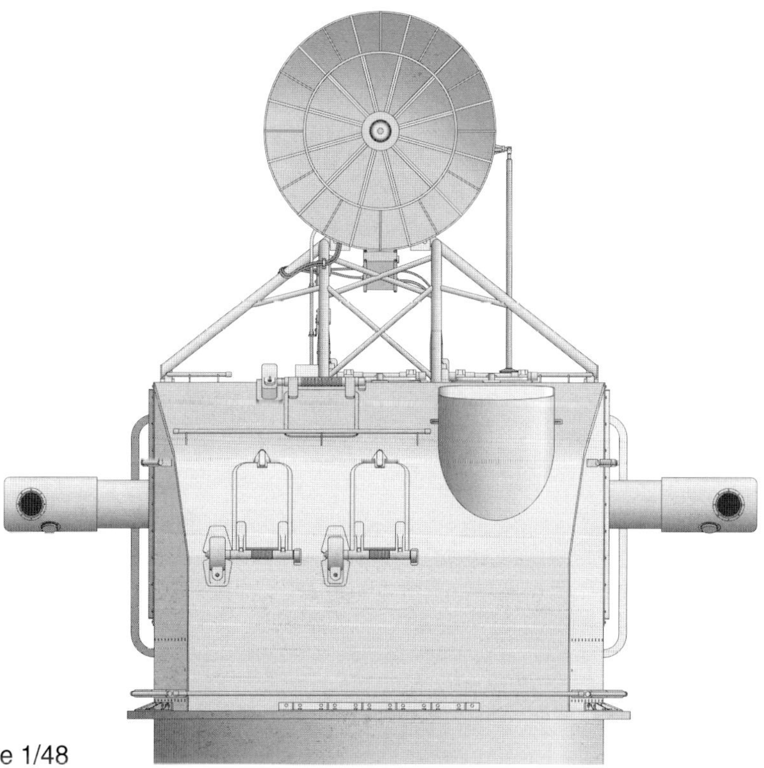

illustration scale 1/48

Two views of Salem *sometime in the early to mid 1950s, in the Mediterranean. Early in her career, she had the midships 3in gun tubs modified, (mounts 35 thru 38) so that they no longer protruded beyond the deck edge. Also, early in her career, she carried six Mk51 directors as back-up 3in fire control. Two were on the bow, four more were amidships (port and starboard, fore and aft of the funnel), all in small tubs. The large round radar antenna atop the mainmast was for the SM-1 radar, which was for air search.*

This photograph was taken around the time of Salem's *completion on May 9, 1949. Three types of directors are visible carried aboard, Mks 37, 54 and 56. For a short time she had a searchlight platform, with signal flag provisions, at the base of the mainmast. The long rectangular basket mounted on the face of the twin 5in turret contained floater nets, which were carried for life-saving purposes. Note the safety netting on the lower portion of the life lines, at the main deck edge.*

May 9, 1949 - The USS Salem CA-139, *is a new ship, about to be commissioned into the greatest navy the world has ever seen. She and her sisters became known as the "Fastest Guns Afloat".*

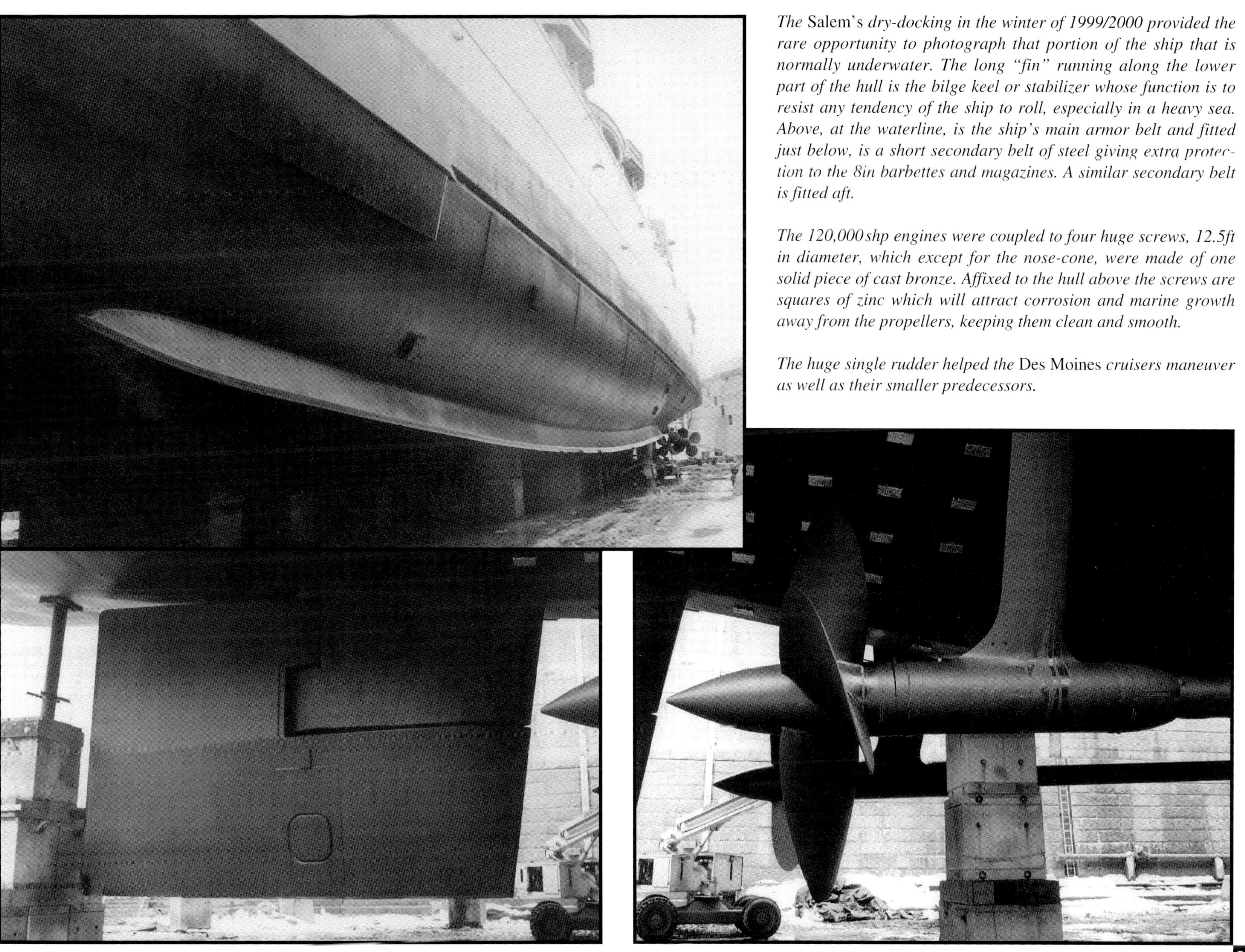

The Salem's *dry-docking in the winter of 1999/2000 provided the rare opportunity to photograph that portion of the ship that is normally underwater. The long "fin" running along the lower part of the hull is the bilge keel or stabilizer whose function is to resist any tendency of the ship to roll, especially in a heavy sea. Above, at the waterline, is the ship's main armor belt and fitted just below, is a short secondary belt of steel giving extra protection to the 8in barbettes and magazines. A similar secondary belt is fitted aft.*

The 120,000shp engines were coupled to four huge screws, 12.5ft in diameter, which except for the nose-cone, were made of one solid piece of cast bronze. Affixed to the hull above the screws are squares of zinc which will attract corrosion and marine growth away from the propellers, keeping them clean and smooth.

The huge single rudder helped the Des Moines *cruisers maneuver as well as their smaller predecessors.*

This is a saluting gun used for ceremonial purposes. It also was used in the high line transfers, firing a line across to the other ship. The range for shooting a line was about 100yds.

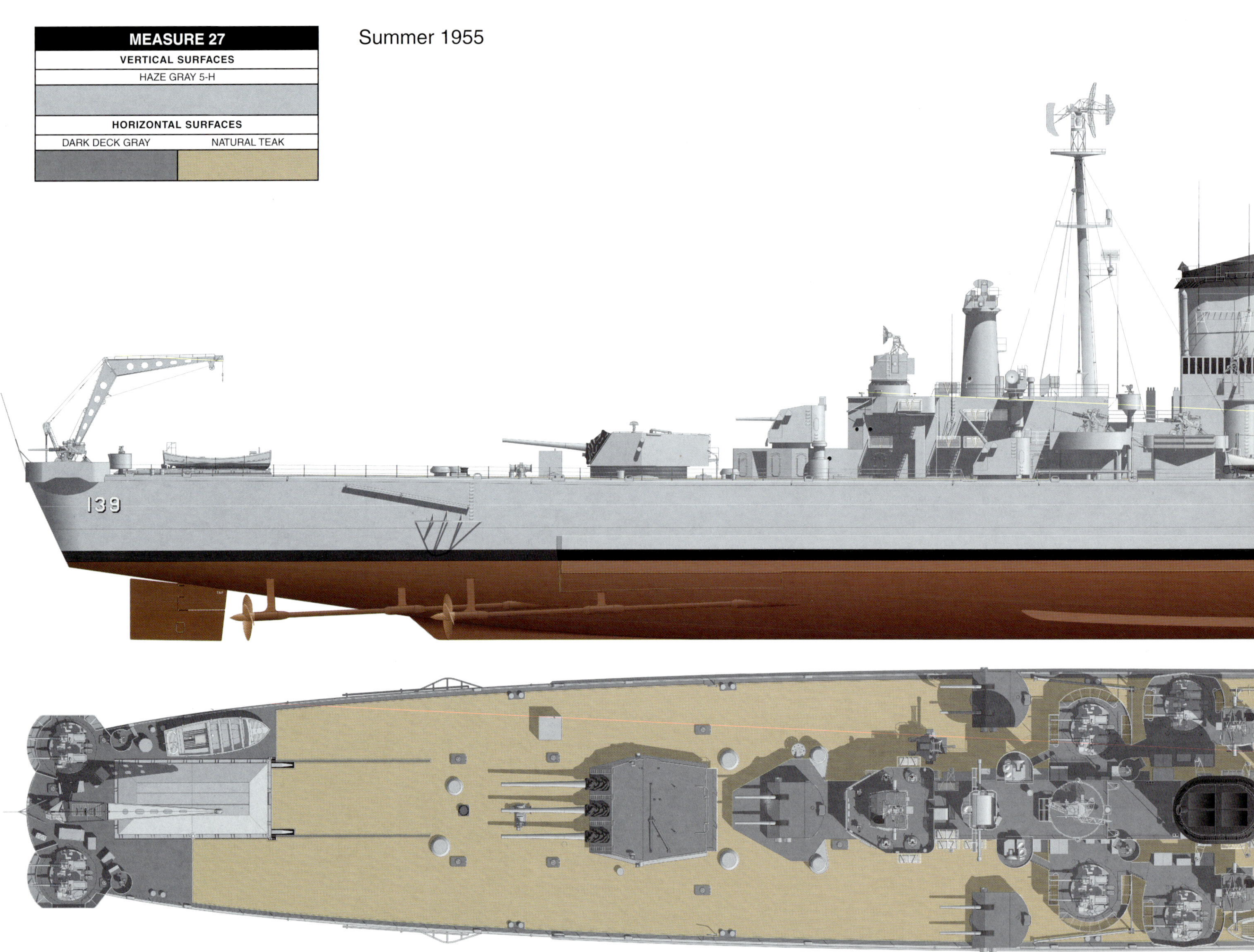
MEASURE 27
VERTICAL SURFACES
HAZE GRAY 5-H
HORIZONTAL SURFACES
DARK DECK GRAY
NATURAL TEAK
Summer 1955
139

USS SALEM CA-139

scale 1/400

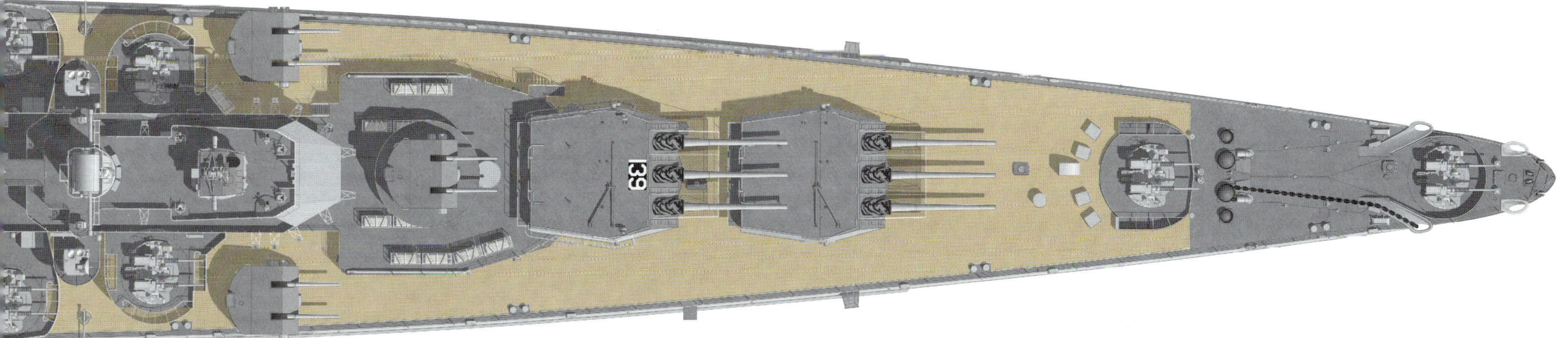

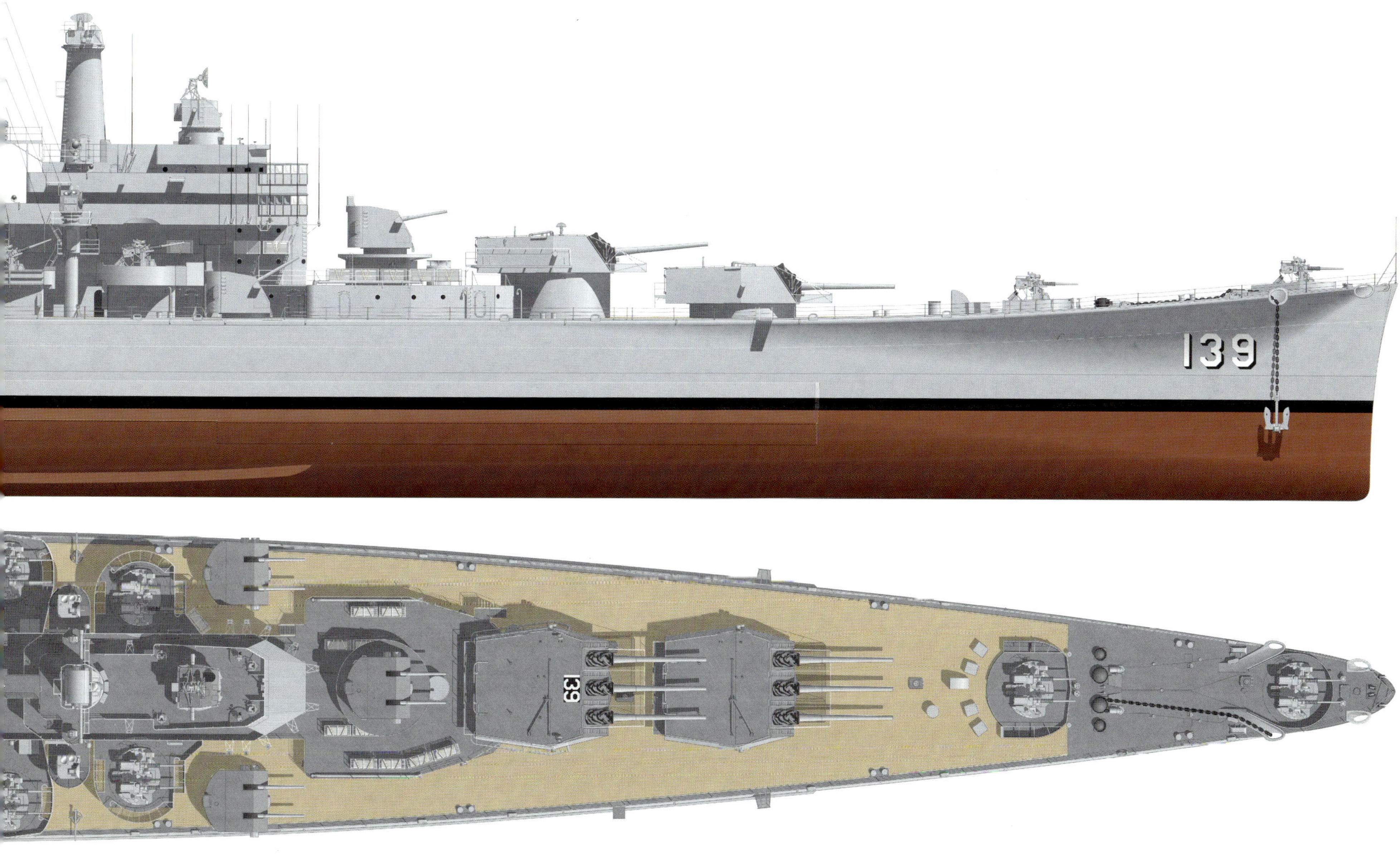

Salem *sometime in 1955, on her sixth deployment to the Mediterranean, to relieve her sister ship, the* USS Newport News CA-148. *She assumed flagship duties during exercises with other vessels under NATO (North Atlantic Treaty Organization). Her painting at this time, and for that matter, her entire career, was Measure 27. All vertical surfaces, from the boot topping (waterline) upwards, to be painted No. 27 Haze Gray (5-H), which was a different Haze Gray than that used during WWII, having a lower content of blue pigment. All horizontal surfaces were painted Dark Deck Gray. All underside and overhead surfaces were painted gloss white. The wood decks remained unpainted (natural teak). All the vertical colors in the postwar paints had a significant increase in gloss, to enhance durability.*

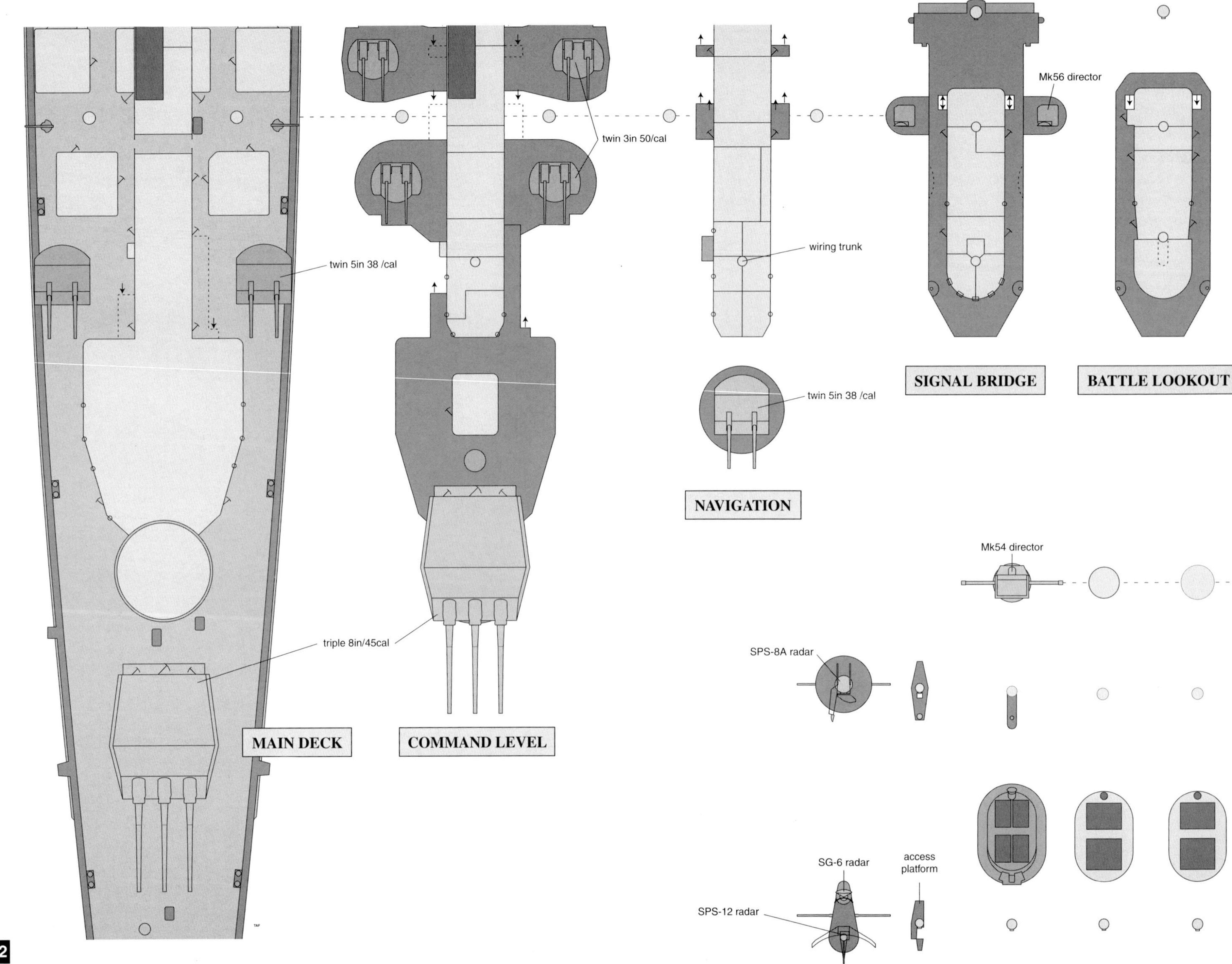

twin 3in 50/cal
Mk56 director
twin 5in 38 /cal
wiring trunk
twin 5in 38 /cal
SIGNAL BRIDGE
BATTLE LOOKOUT
NAVIGATION
Mk54 director
triple 8in/45cal
SPS-8A radar
MAIN DECK
COMMAND LEVEL
SG-6 radar
access platform
SPS-12 radar

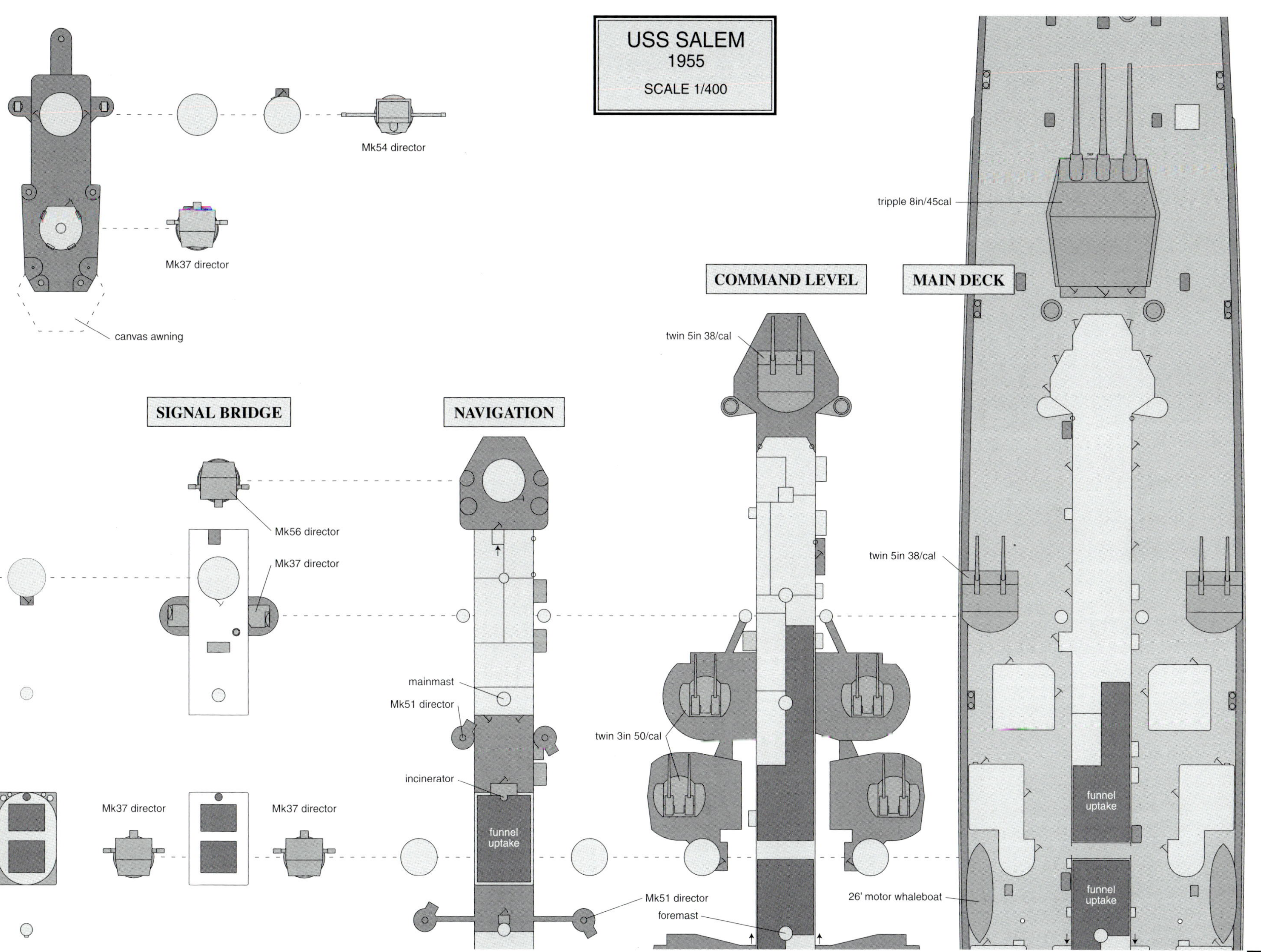
USS SALEM
1955
SCALE 1/400
Mk54 director
Mk37 director
canvas awning
COMMAND LEVEL
MAIN DECK
tripple 8in/45cal
twin 5in 38/cal
SIGNAL BRIDGE
NAVIGATION
Mk56 director
Mk37 director
twin 5in 38/cal
mainmast
Mk51 director
twin 3in 50/cal
incinerator
funnel uptake
Mk37 director
Mk37 director
funnel uptake
funnel uptake
Mk51 director
foremast
26' motor whaleboat

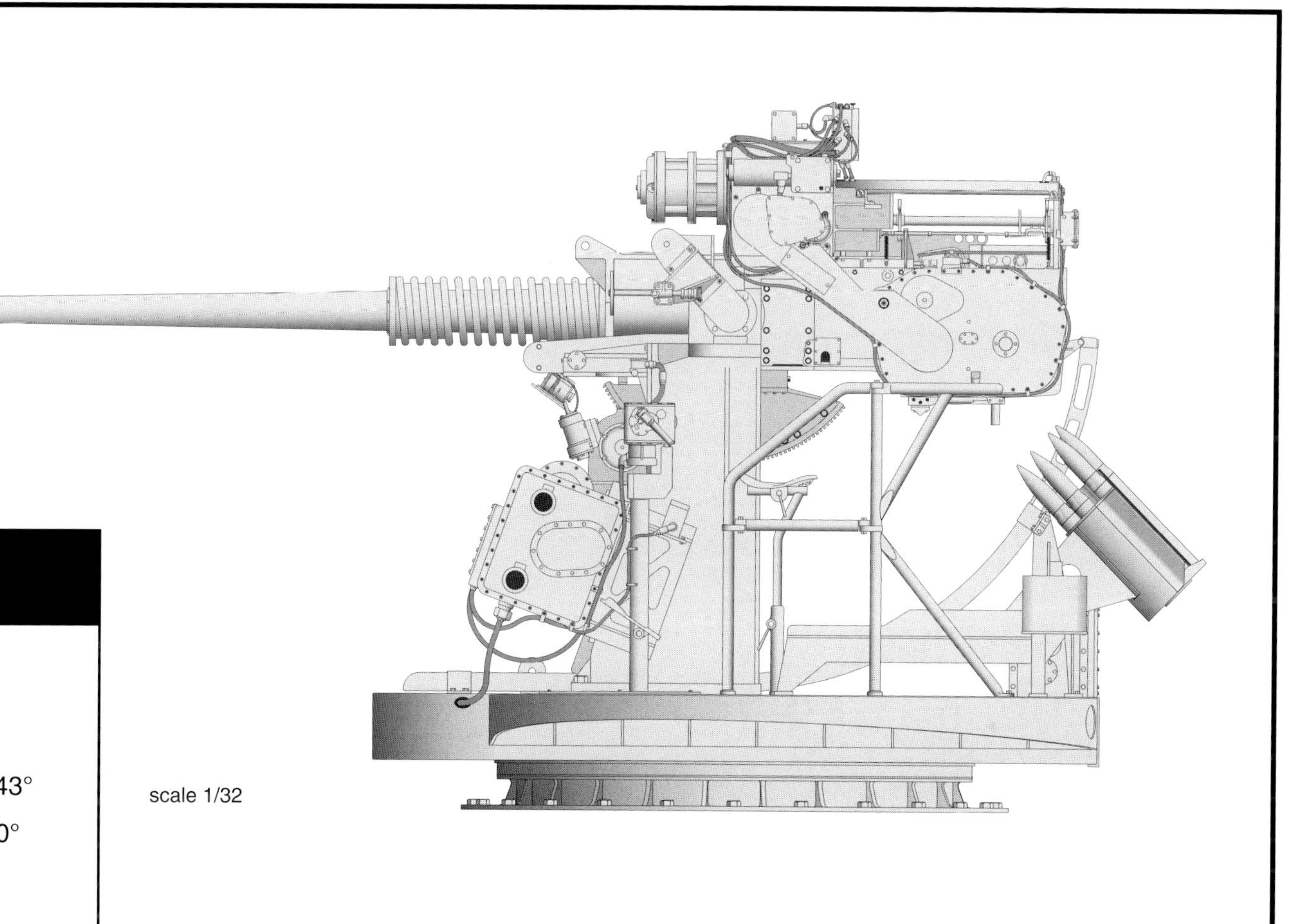
scale 1/32

3 in ANTI-AIRCRAFT

50 caliber Mk 22

Muzzle velocity	2,700 fps
Maximum range	14,600 yd @ 43°
Maximum ceiling	29,800 ft @ 80°
Rate of Fire	50 rpm
Approx life	4,300 shells

Mount

Mk 27 & 33

Weight	16.5 tons
Gun elevation	85° to -15°
Elevation speed	15° ps
Training speed	30° ps
Crew	12

The 3 in/50cal automatic gun was developed in direct response to the Japanese kamikaze attacks on Allied warships during the later part of World War II. A kamikaze pilot was the expendable guidance system in an aircraft which was packed with explosives and was basically a guided missile. It was necessary to literally destroy the aircraft as far away from its target as possible. Developed from the older 3 in/50 into a auto loading weapon, the new gun was able to fire four times faster. It was fully auto directed locally on some mounts, by Mk 34 radar or remotely, on most mounts, by the Mk 56 director. When mounted in pairs, as was most often the case, the twin Mk 27 could fire 100 proximity fused rounds at an enemy aircraft before it was within two miles of the ship. Air defense weapons of the future used a similar approach to combat incoming enemy missiles until replaced with anti-missile missiles or Gatling type automatic guns.

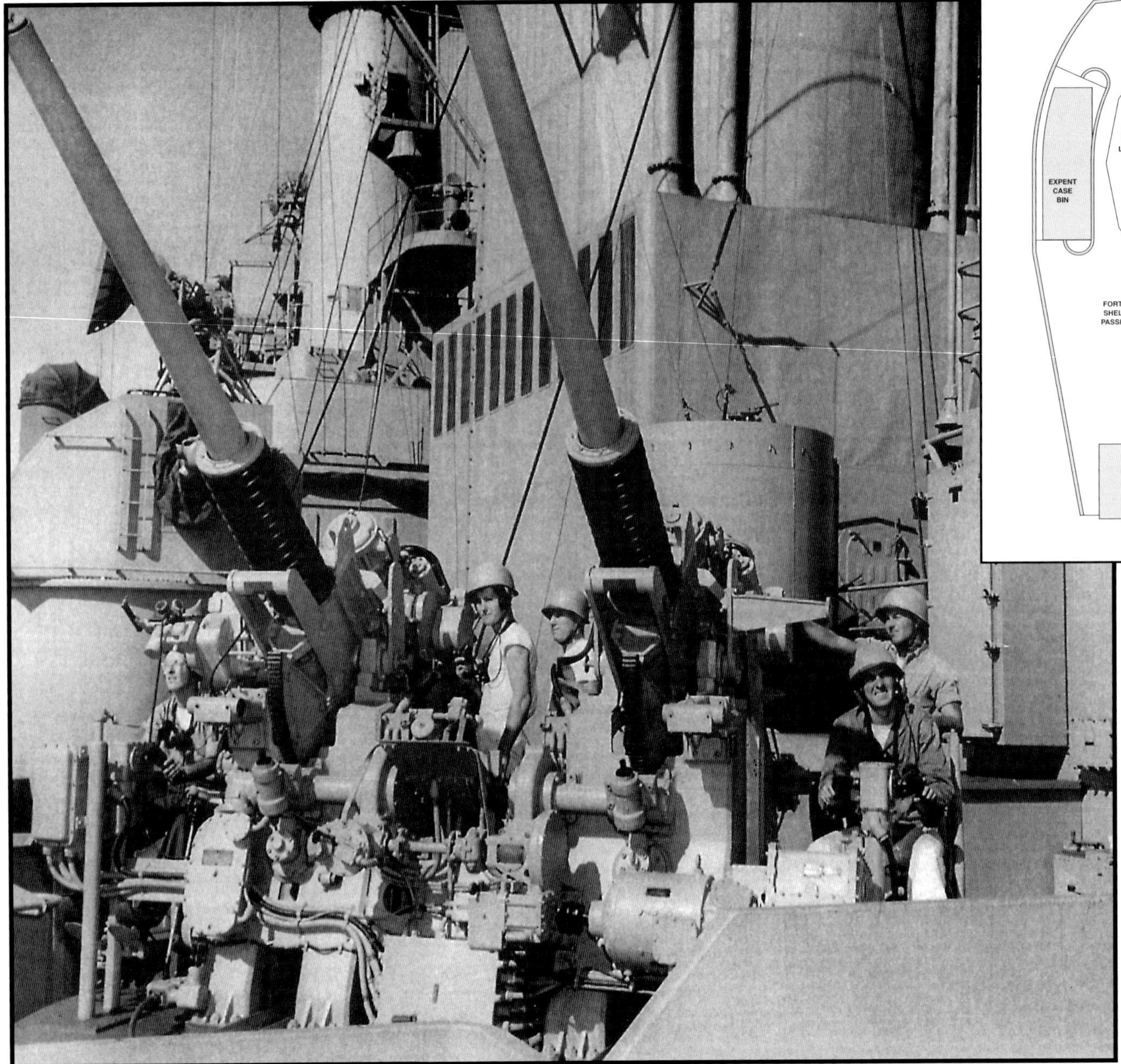

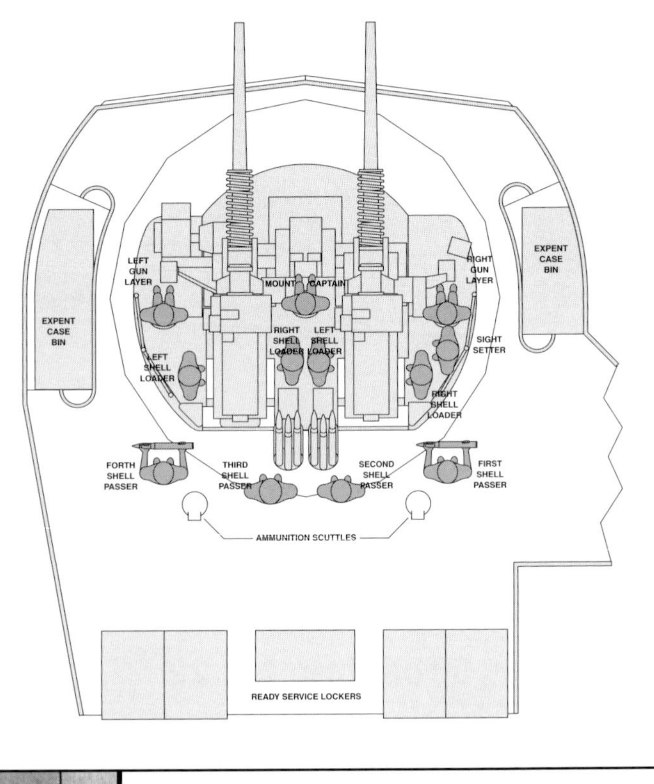

The 3in twin-auto was operated by a crew of 12, which generally consisted of 11 enlisted men and one officer, the mount captain. Although the 3in auto was self-loading, the gun's magazine was hand stocked and the weapon's high rate of fire ment that magazine loading was fast and furious. Shells were stored in near-by lockers and additional rounds were received through ammunition scuttles.

Mk56 - 3in gun director

Developed at MIT University, the Mk56 was the principal postwar director for the 3in twin automatic gun. It was suited for blind firing at surface as well as air targets. The director centered around the Mk35 radar to incorporate auto tracking in range, bearing and elevation. Effective distance was 30,000yds on a bomber at speeds up to 630 miles per hour. A sluing sight gave the operator a 30° lead angle on moving targets. The entire director weighed 9.6 tons. Above deck, the director proper carried a crew of two with four additional men below deck to operate the ballistic computers. *Salem* carried a total of four Mk56 directors.

scale 1/32

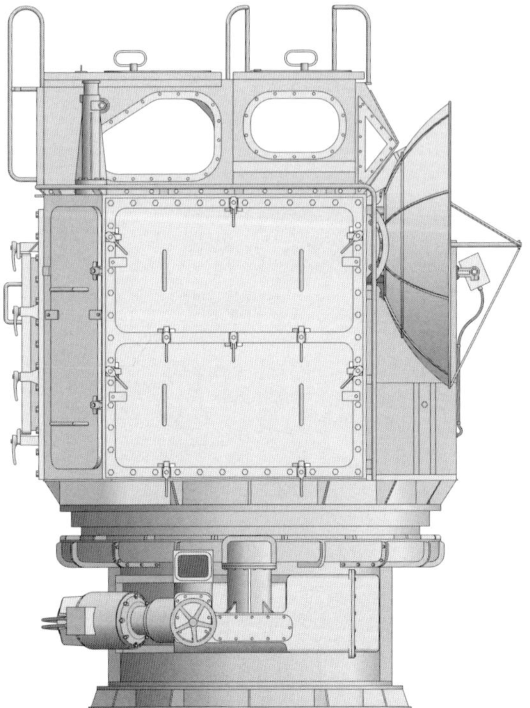

scale 1/32

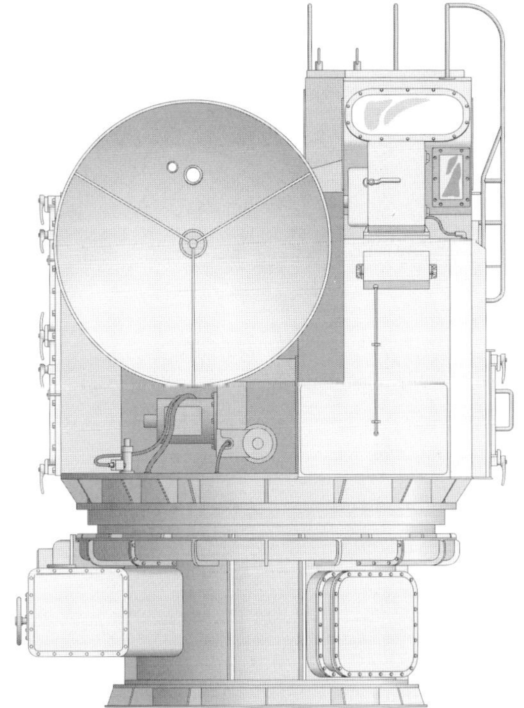

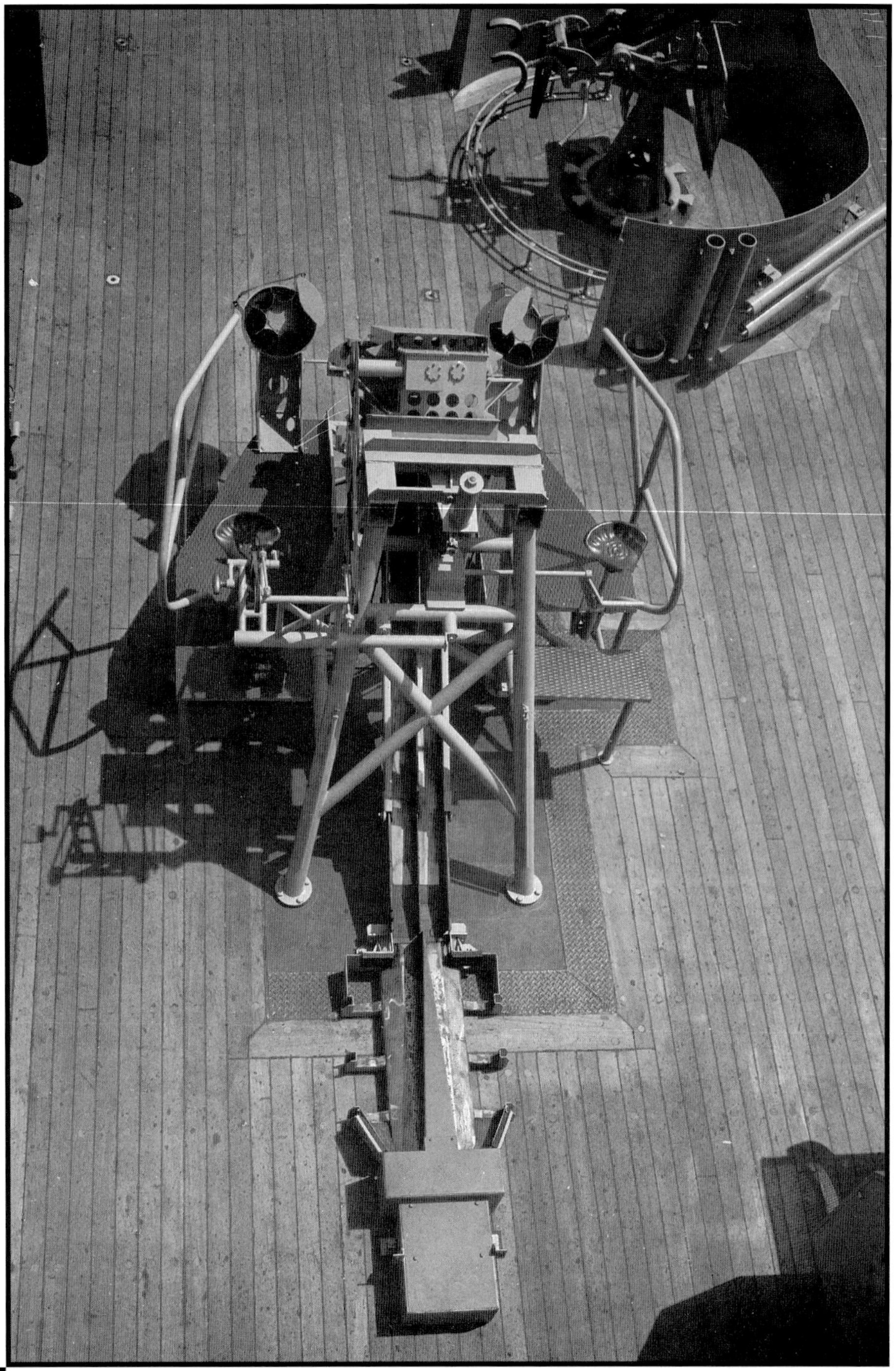

Although the 3in anti-aircraft gun was considered an automated weapon, the gun's magazine was hand loaded. The ammunition stored nearby in lockers was transferred by "shell passers" to the rear of the 3in mount. "Shell loaders then fitted the ammunition into the auto-loader mounted at the rear of each gun. To maintain a continuous feed and uninterrupted fire, a certain amount of practice was needed. The Salem *had one practice mount fitted amidships solely for the purpose of keeping the loading personnel sharp. Practice secessions were held on a regular basis and contests decided which gun crews received extra practice time. Dummy shells were used and all of the wear and tear on machinery was confined to the practice mount. This machine was in the shape of the single mount, even though* Salem *had all twin mounts.*

Visible at the rear of the practice loader, sits one of the twin 20mm mounts. When built, the Salem *had six of these light anti-aircraft weapons fitted, but they were removed soon after, as faster, heavier aircraft rendered these weapons obsolete. Two vertical tubes attached to the shielding were used for storing overheated 20mm gun barrels while they cooled. The angled tubes held the replacement gun barrels .*

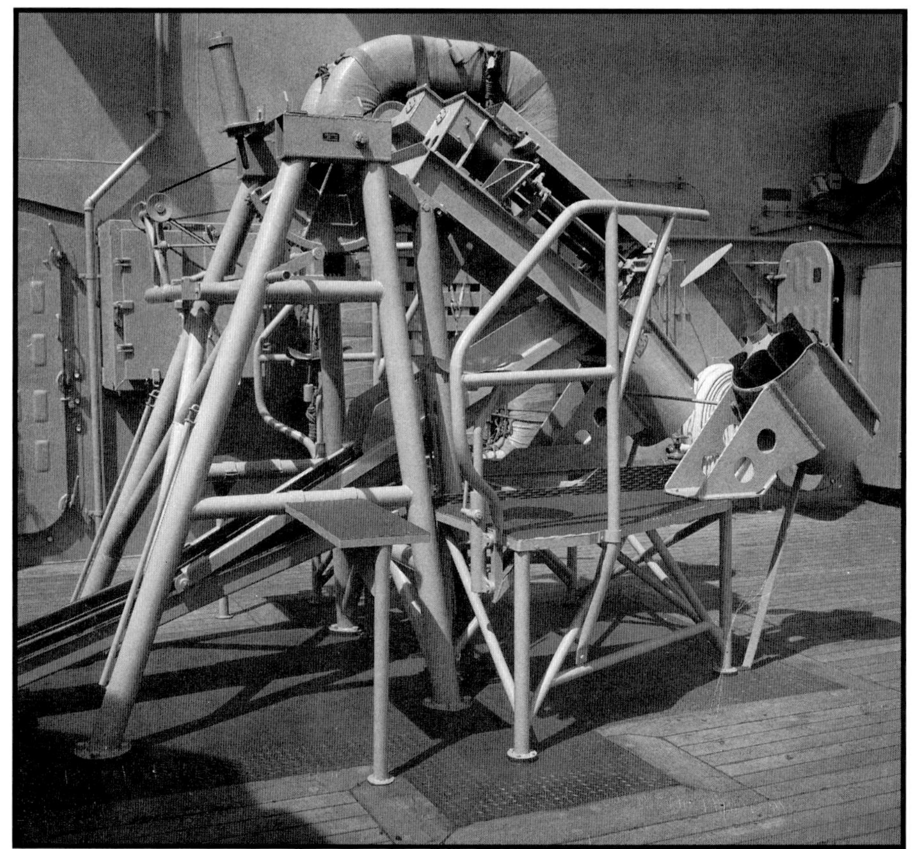

USS Salem, *sometime in the mid 1950s, probably in 1954 or 1955. She has had her radar upgraded to the SPS-12 and SPS-8, but still retains mount 31 in her twin 3in. anti-aircraft battery. The independent Mk51 directors for that mount, and number 32 have been deleted. The four such directors still remain amidships, abreast the funnel. A little known fact about the aircraft crane on the stern, they were basically the same design since their inception when they were first fitted on battleships in the mid to late 1930s. As originally designed, this class of cruisers, were to have catapults, and spotter/scout aircraft. By the time of her building,* Salem *did not have the catapults installed, but did have the hangar built into the stern. This was used for repair and maintenance of the helicopters that were aboard during the 1950s.*

The USS Salem's *design included the latest electronic equipment and radar systems. This included primary and secondary communications, radio and radar interruption devises, surface and air search radar and fire control radar. During a career of about ten years, the* Salem *underwent numerous upgrades and improvements to maintain the most up-to-date fighting capability possible. The ship's final configuration included the primary search radar units featured in illustrations on page 43.*

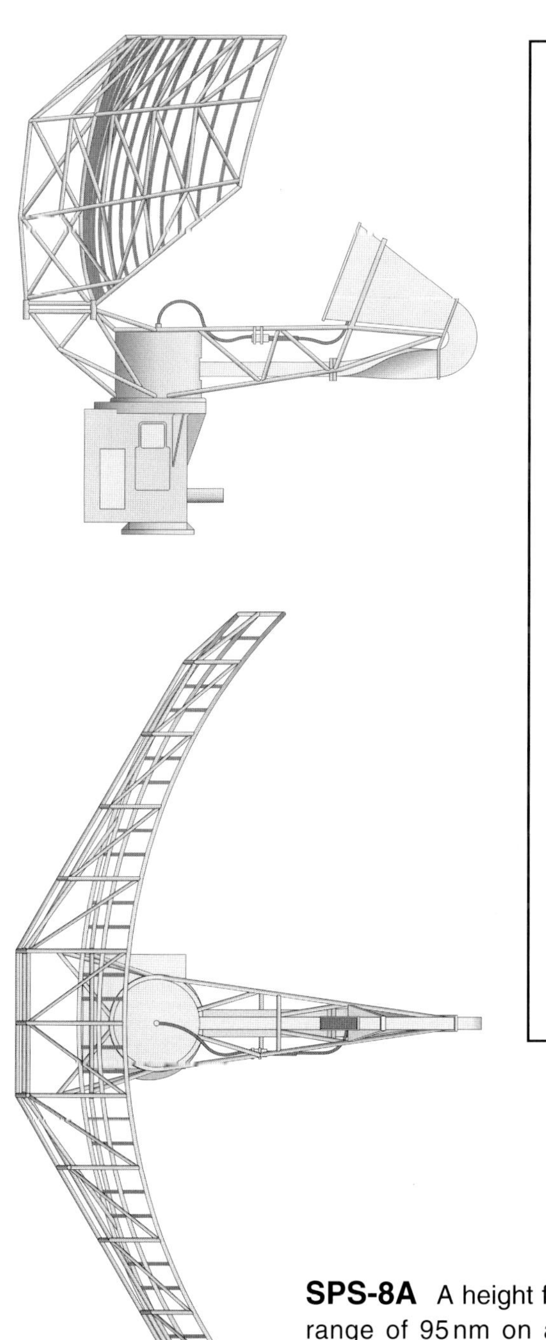

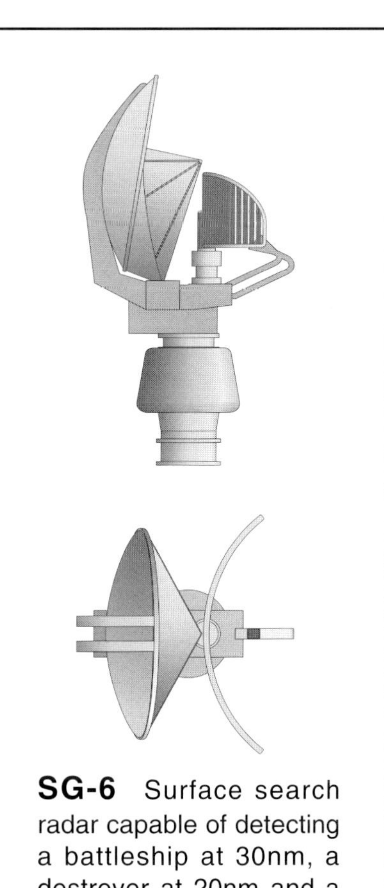

SG-6 Surface search radar capable of detecting a battleship at 30nm, a destroyer at 20nm and a surfaced submarine at 12nm. Made by the Raytheon Corporation.

illustrations scale 1/48

SPS-8A A height finding, air search radar having a range of 95nm on an aircraft at 10,000ft. Aircraft could be detected down to a height 500ft, or as far away as 160nm. This type was in use from 1952 thru the mid 1960s. Made by General Electric.

SPS-12A Air and surface search radar capable of detecting aircraft at 90nm and a ship at 50nm. This type was in use from 1954, thru the 1970's. Made by Westinghouse and Bendix.

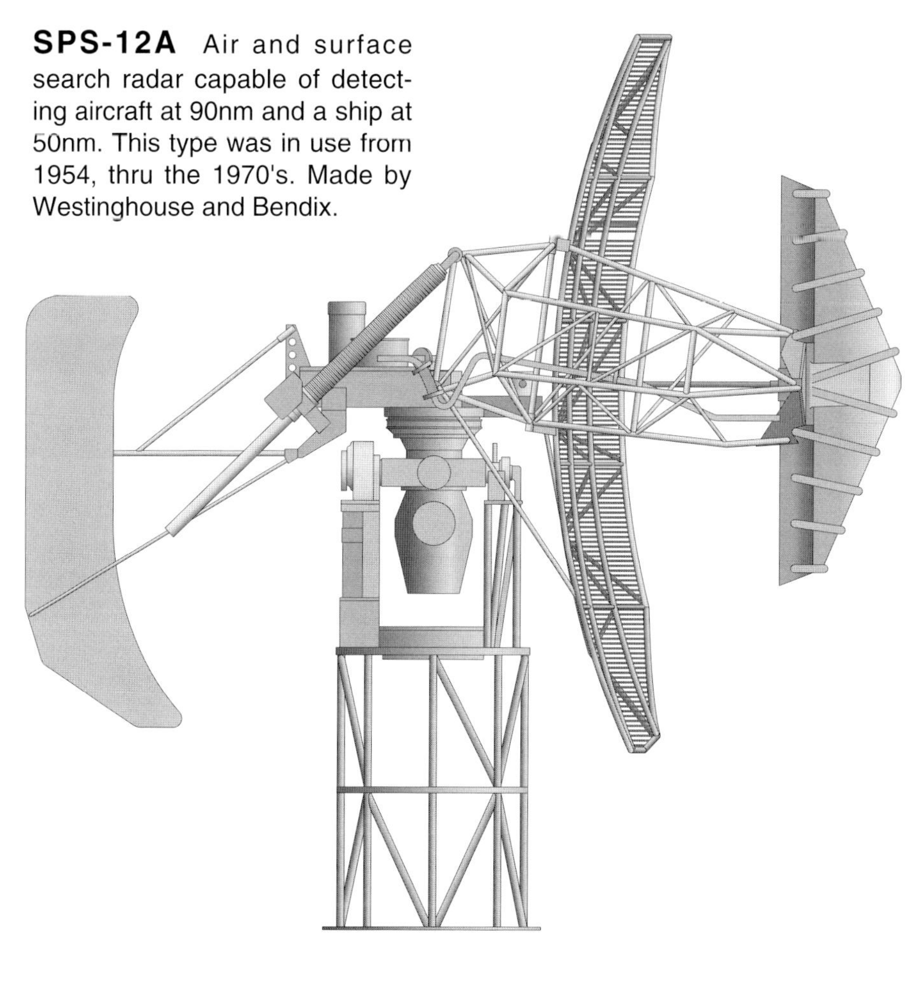

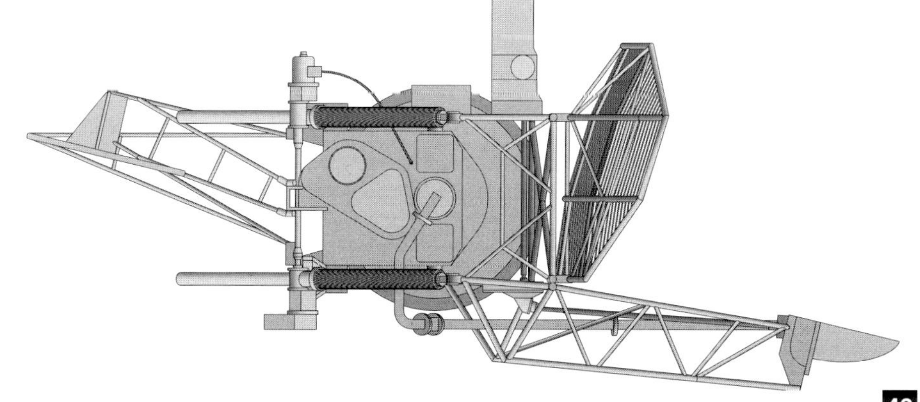

USS Salem, *upon completion, May 9, 1949*

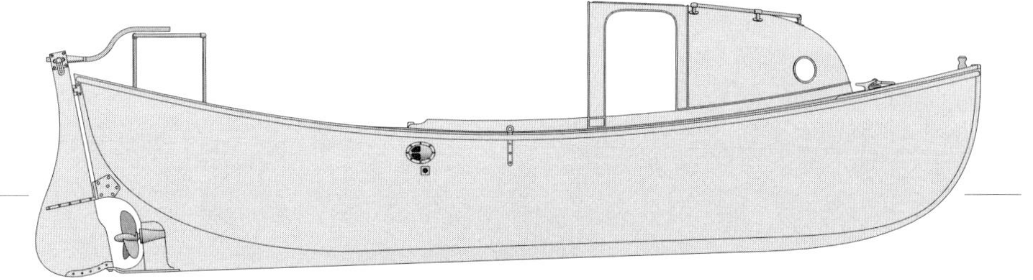

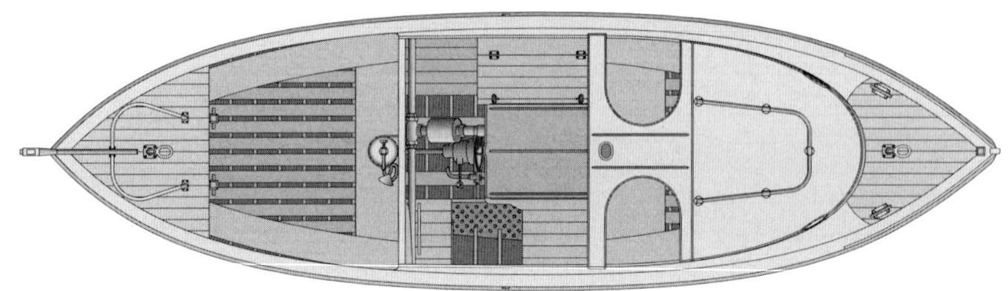

26ft motor whale boat

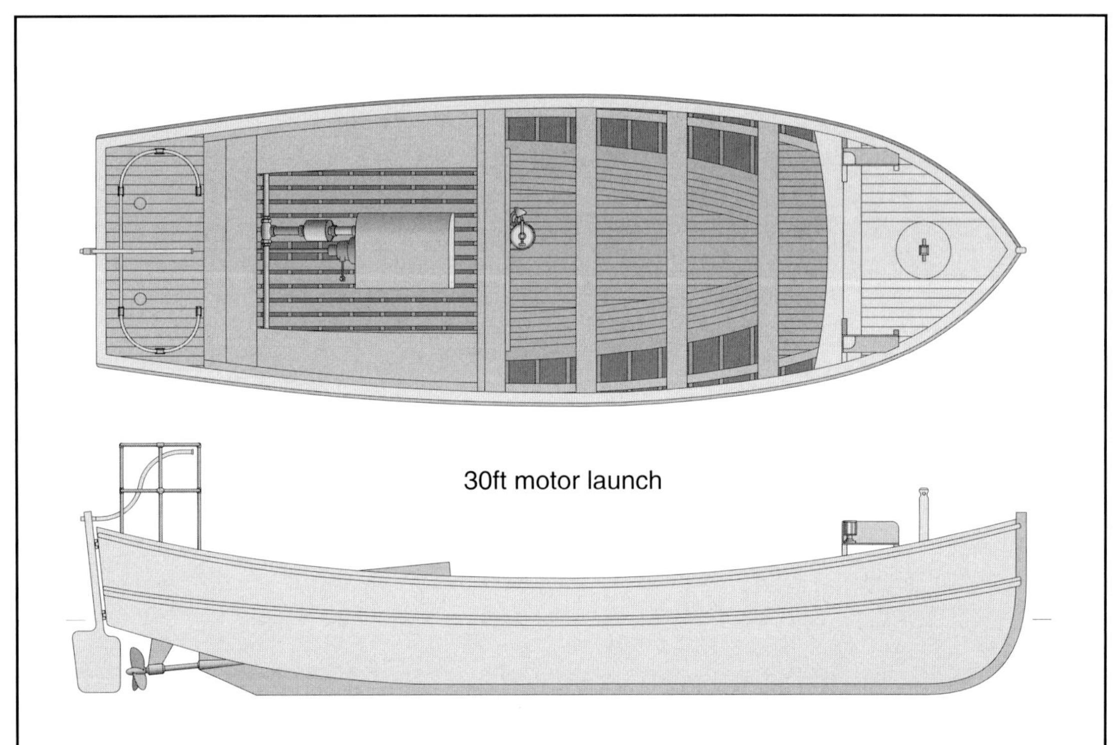

30ft motor launch

SHIP'S BOATS

The USS Salem carried a complement of small craft to ferry material/men to and from shore when the ship was not docked. The 28 ft Personnel was probably reserved for use by officers.

26 ft Motor Whaleboat

Capacity	22 including 2 crew
Displacement	9,000 lb full load
Engine	1 - 4cylinder diesel
Speed	7 kt at full load
Fuel capacity	28 gallons
Range	110 nautical miles

30 ft Motor Launch

Capacity	40 including 2 crew
Displacement	15,000 lb full load
Engine	1 - 4cylinder diesel
Speed	6 kt at full load
Fuel capacity	50 gallons
Range	160 nautical miles

scale for illustrations on pages 46/47 1/48

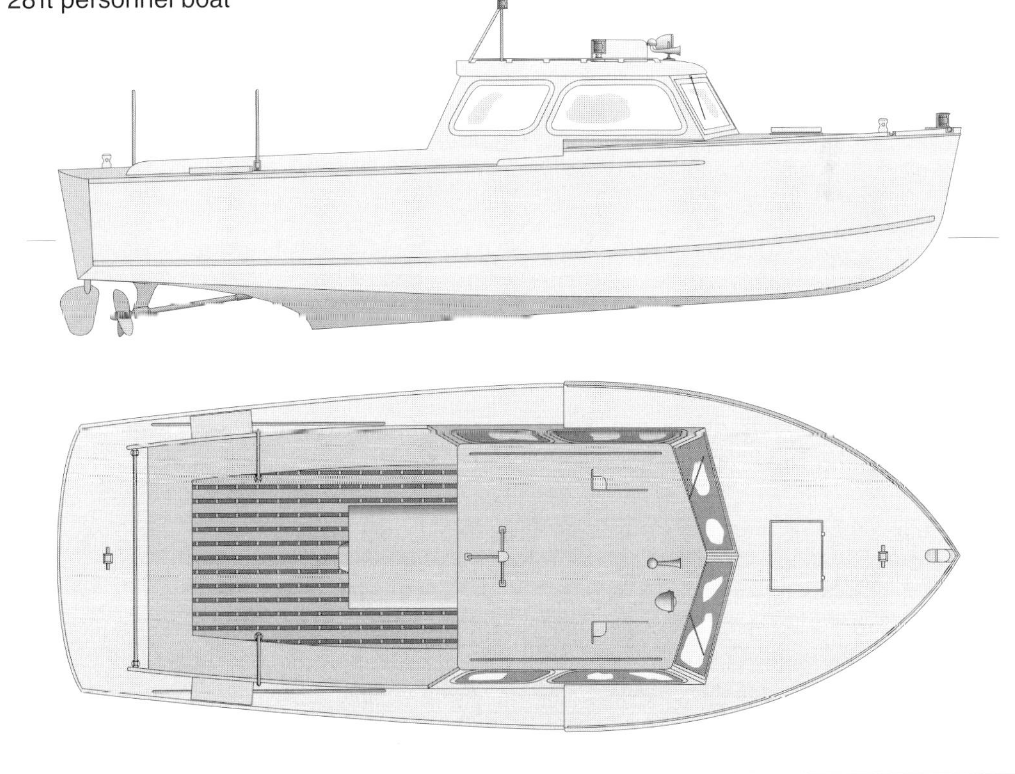

28 ft Personnel Boat

Capacity	22 including 2 crew
Displacement	12,600 full load
Engine	1 - 6 cylinder diesel
Speed	16 knots @ full load
Fuel Capacity	90 gallons
Range	120 nautical miles

The floater net system and 25 man life rafts were the ship's original equipment for emergency flotation. Both were discarded in favor of the self-inflating life raft which was an altogether better means of survival in the water as well as being very compact when not in use.

Floater nets were convenient, but of only short term value in use.

25 man life rafts strapped to the side of # 3 8in turret.

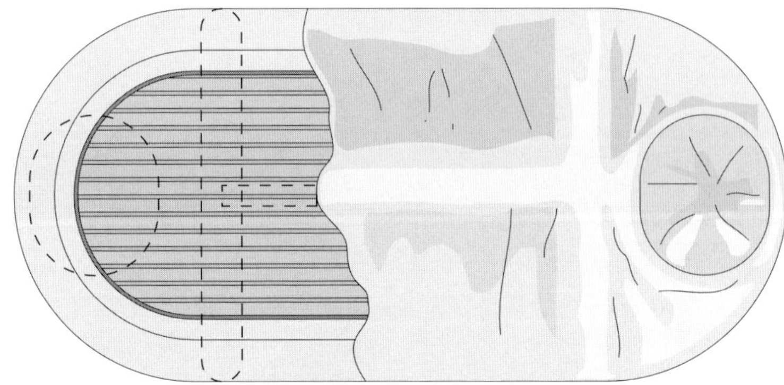

co/2 life raft illustration 1/48

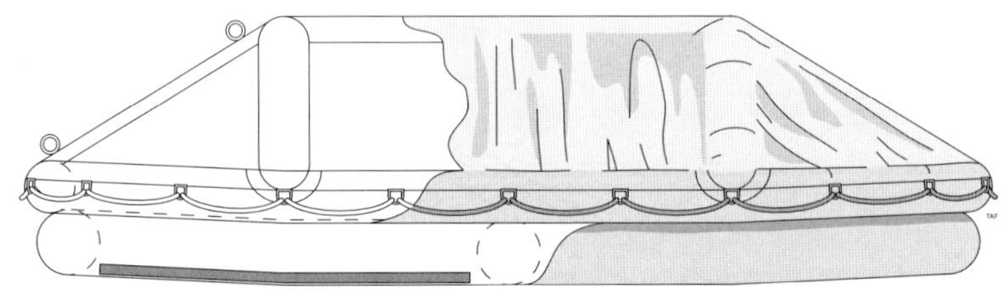

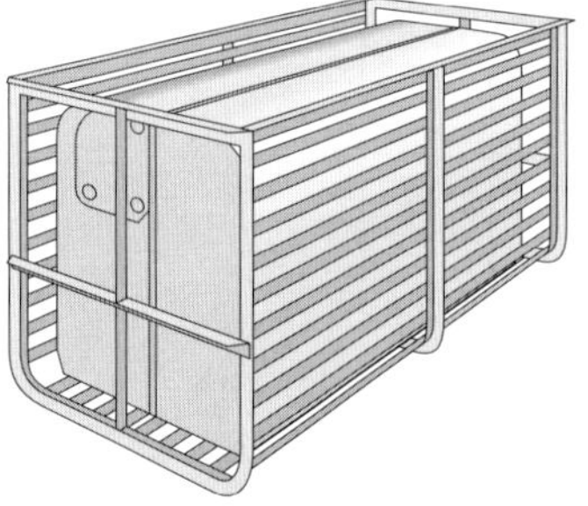

co/2 life raft stored in it's container basket.

The inflatable life raft carried by the Salem *was rated for 15 men and weighed about 225lb. Stocked with normal provisions, the raft weighed about 350lb. When needed, the raft was released from the basket, unfolded and inflated with co/2 from the self-contained tank. Provisions were then loaded and the raft thrown overboard. Survivors would enter the raft from the water. A built-in overhead cover kept the sun at bay.*

The "Sea Witch" *late in her career, about 1957 or 1958. She has had the forward 3in/50cal twin mount, #31, removed. This mount required constant servicing due to damage from the seas coming over the bow. She wears her final complement of radio and radar antenna on both the superstructure and the mastwork. Note how the mainmast is painted gloss black down to the base level of the funnel cap, and the foremast is not painted black at all.*

Sikorsky HO3S-1 (H-5)

Designed from a prototype that first flew in 1943, they were an upgraded version that had a 600hp radial engine. A total of 90 of these helicopters were delivered to the US Navy before production was ended in 1951. They would serve into the mid 1950's aboard ships as rescue and ship to ship, or shore transportation when one of the ships boats just was not practical. They had a four seat cabin. Their fuselage measured to a length of just over 42 feet. The diameter of the main rotor was 48 feet. Maximum speed was 104mph, service ceiling at 14,000 feet, and a range of 300 miles. The aircraft was painted in an overall glossy Sea Blue, with Dayglo Orange highlighted danger zones.

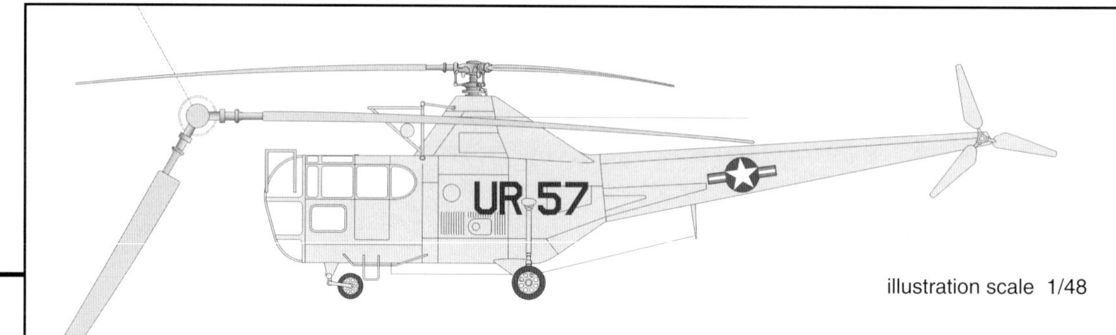

illustration scale 1/48

A very powerful view of Salem *in the Mediterranean, late in her career, probably during her 20 month cruise as Flagship of the US 6th Fleet. This photograph can be dated as such because of certain changes. The bow 3in twin gun has been removed. Both the SPS-12 and SPS-8 radar are visible on the fore and mainmast, respectively. This is also a good view of the hull form, which proved to be a successful design, due to the high, wide, and well flared bow shape. Her high freeboard also proved to be an asset on the open ocean, keeping her dryer than previous classes of cruisers.*

Salem's *sister ship and the name ship of the class,* USS Des Moines CA-134. *She was built at the same shipyard as the* Salem, *having been laid down on May 28, 1945, launched on September 27, 1946, and commissioned on November 16, 1948. She was the only one of the class to have been built with catapults, which were removed soon after entering service. Like the* Salem, the Des Moines *would serve in the Mediterranean with the US 6th Fleet for the majority of her career.* CA-134 *was decommissioned on July 14, 1961. She was stricken from the USN roster on July 9, 1991, and lies in reserve at the Philadelphia Naval Shipyard at the date of this printing.*

The Salem's *other sister,* USS Newport News CA-148. *She was built at the Newport News Shipyard, with the keel being laid on October 1, 1945, launched on March 6, 1948, and commissioned on January 29, 1949. Like her sisters,* CA-148 *was part of the US 6th Fleet and operated in the Mediterranean for all of the 1950s. After a major refit in the early 1960s,* Newport News *operated in the Atlantic and Caribbean Oceans, eventually going to the Pacific in 1967. She was the only one of her class to venture into that ocean. She was also the only one of the class, according to official records, to have fired her guns in anger, during the Vietnam War.* CA-148 *made two tours during that conflict. The* Newport News *has the distinction of being the last heavy cruiser in commission in the US Navy, and the world's navies, until her decommissioning on June 27, 1975. She was stricken on July 31, 1978. The* Newport News *was sold to the Southern Scrap Metals Corporation at New Orleans, Louisiana, and broken up from 1993, into 1995.*

Two stern views of the Salem *while in the Constitution Dock at the Boston Ship Repair Yard, during her dry-docking in late 1999, into early 2000. The overhang of 3in mounts numbers 311 and 312 is shown here quite well. Note how they are flared into the hull from underneath. The plating detail on the stern area of the hull is also visible from these photographs. The two items inset into the stern hull, up near the deck edge, are the stern lights. As large a ship as the* Salem *is, the dry dock still has plenty of room to spare. This is possibly the largest of its kind, still in use,on the upper east coast of the United States.*

USS Salem CA-139 *in the "Mothball Fleet" at the Philadelphia Naval Shipyard, 1967. The ship has been completely weatherproofed. The gun barrels have all been sealed. All doors and hatches on the superstructure have been sealed. The entire ship is sealed to preserve her, if needed, for active duty at a later date. Note the sealed bridge windows and the domes enclosing the twin 3in mounts. Often, when these ships are put into reserve, weapons, radar, and all other sorts of equipment, were removed and used where needed on other ships. From what little we can see here, the radar antenna on both the Mk37 and 54 directors have been removed. The rangefinder in the Mk54 director has been removed. It would also appear that the SPS-12 radar antenna has been removed. The "Bloomers" have also been removed from the main gun turret.*

Directions to the USS Salem-

<u>United States Naval & Shipbuilding Museum</u>
<u>USS Salem CA-139</u>

739 Washington Street
Quincy, Massachusetts 02169
Phone (617) 479-7900 Fax (617) 479-8792

The USS Salem and United States Naval & Shipbuilding Museum are located in the city of Quincy, MA., just south of Boston, off of Route 3A, next to the Fore River Bridge.

Museum Hours

Daily, 10:00AM - 7:00PM

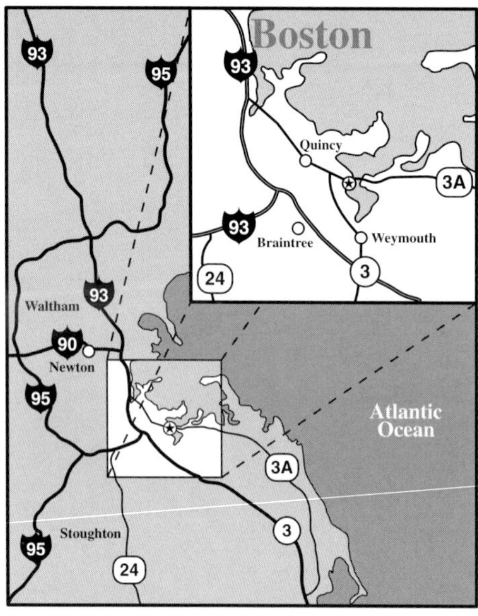

Front Cover Art
by

Tom Freeman

S M & S Naval Prints
P. O. Box 41
Forest Hills, Maryland 21050-0041
Phone (410) 893-8184
Fax (410) 879-1261

Two more views during the Salem's *dry-docking in Boston, in the winter of 1999/2000.*

GENERAL STATISTICS

Ordered
June 14, 1943

Builder
Bethlehem Shipbuilding Corporation
Quincy Massachusetts

Keel Laid
July 4, 1945

Launched
March 25, 1947

Commissioned
May 14, 1949

Dimensions (ft)

length overall		716.50
length waterline		700.00
beam		76.35
draught	(min.)	25
	(max.)	26.00
height	(above wl)	142.85

Displacement (tons)

light	15,869
standard	17,273
normal	20,934
full load	21,493

Boilers
Babcock & Wilcox (4)

Engines
General Electric geared turbines (4)

Speed..........32.5 kts at 120,000 shp

Fuel (tons)

standard	2675
full load	3007

Endurance
10,500 nm at 15 kts
8,050 nm at 20 kts

Armament

main battery
8 in/55 cal • Mk 16
3 turrets x 3
director Mk 54

secondary battery-anti-aircraft
5 in/38 cal • Mk 12
6 mounts x 2
director Mk 37

anti-aircraft
3 in 50 cal • Mk 34
12 mounts x 2
director Mk 56 & Mk 63

20 mm • Mk 26
6 mounts x 2
director Mk 51

(all 20 mm guns and their directors were removed by 1952)

Armor (in)

main belt		4 to 6
decks	(bomb)	1
	(armor)	3.5
turrets		2 to 8
barbettes		6.3
secondary turrets		2 to 3.75
bulkheads		5
conning tower		4 to 6.5

(armor weight comprised 13.8% of light ship)

Aircraft
Sikorsky HO3S-1 1

Small craft

26 ft Motor Whale Boat	2
28 ft Personnel Launch	1
30 ft Motor Launch	1

Rescue

early	life rafts (25 man)	16
	floater nets	36
late	inflatable life rafts (15 man)	65

Complement
109 officers • 1690 enlisted men

Radar Systems

target acquisition

surface search		SG-6
air search	(early)	SPS-6
	(late)	SPS-12
height finding		SPS-8A
fighter control	(1949/1953)	SP

weapon control

8 in/55 cal	Mk 13 & Mk 27
5 in/38 cal	Mk 25
3 in/50 cal	Mk 34 & Mk 35

Decommissioned
January 30, 1959

Cost to build
$48.1 million (1949)

REFERENCES

Cruisers of WWII
Naval Institute Press, 1995, M.J. Whitley

Dictionary of American Naval Fighting Ships
US Government Printing Office, 1990

U.S.Cruisers
Naval Institute Press, 1984, Norman Friedman

U.S. Naval Weapons
Naval Institute Press, 1985, Norman Friedman

Naval Radar
Conway Maritime Press, 1988, Norman Friedman

Warship International
Naval Records Club, Vol. IX, #4, 1972, E. C. Fisher Jr.

RESOURCES

Real War Photos
P.O.Box 728, Hammond, IN 46325

The Floating Drydock
P.O.Box 250, Kresgeville, PA 18333

U.S.Naval Historical Center
Building 57
Washington Navy Yard, Washington D.C. 20374-2571

U.S.Naval Institute
291 Wood Rd, Annapolis, MD 21402-5035

CLASSIC WARSHIPS PUBLISHING
extends a very special thanks to

Robert Duetsch, James Fahey & John Frohock of the
United States Naval & Shipbuilding Museum
USS Salem CA-139

Tom Freeman, A. D. Baker III, Louis Parker,
Don Montgomery, & John Sheridan

On the back • USS Salem CA-139 *at anchor in the Canal Della Guidecca, during a visit to Venice, Italy, August 23, 1951.*

Listed below are some of our favorite sources for reference books, plans photographs, and models

REAL WAR PHOTOS
P.O.Box 728
Hammond, Indiana 46325
catalog $3

PACIFIC FRONT HOBBIES
P.O. Box 2098
Roseburg, Oregon 97470-2098
Ph. 541-464-8579
catalog $5

THE FLOATING DRYDOCK
P.O. Box 250
Kresgeville, Pennsylvania 18333
catalog $5

U. S. NAVAL INSTITUTE
291 Woods Road
Annapolis, Maryland 21402
Ph. 800-233-8764

TAUBMAN PLANS SERVICE
11 College Drive #4
Jersey City, New Jersey 07305
catalog $10